Nanoagronomy

Sumera Javad
Editor

Nanoagronomy

Springer

Editor
Sumera Javad
Department of Botany
Lahore College for Women University
Lahore, Pakistan

ISBN 978-3-030-41274-6 ISBN 978-3-030-41275-3 (eBook)
https://doi.org/10.1007/978-3-030-41275-3

This Springer imprint is published by the registered company Springer Nature Switzerland AG
The registered company address is: Gewerbestrasse 11, 6330 Cham, Switzerland

Preface

This book is an effort to establish the role of nanotechnology in the field of agriculture. Nanoparticles have a broad spectrum of applications, and one of them is deeply rooted in agriculture. Successful agriculture is a guarantee to continuous supply of food for the ever-increasing population. As the world is moving towards the scarcity of food and agricultural resources, there is a real need of searching for better and alternative smart techniques for applied agriculture and here nanotechnology seems to be playing its role. In this book, we have tried our best as a team of authors to present the associated problems of agriculture and proposed solutions from nanotechnology, which is going to be/define the future of agriculture. It will be a good source of knowledge for students of agriculture, agronomy, and plant sciences.

Lahore, Pakistan Sumera Javad

Contents

About the Editor

Sumera Javad is an Assistant Professor in the Department of Botany at Lahore College for Women University, Lahore, Pakistan. She did her PhD in Botany from Pakistan with specialization in plant tissue culture and plant metabolites. She got her postdoctoral research experience from Food Science Department, Cornell University, NY, USA, under the Postdoctoral fellowship of AAUW (American Association of University Women), Washington DC, USA. She is working especially on plant-based bio-chemicals and nanotechnology related to plants. She also served as editor of *Nanobotany* published by Springer. She has also published a number of international articles.

Chapter 1
Nanotechnology: A Breakthrough in Agronomy

Madeeha Ansari, Kiran Shahzadi, and Shakil Ahmed

1.1 Introduction

Farming is the practical implementation of agriculture industry (Fig. 1.1), the process in which food, feed and fibers are produced. One of the most important branches of agricultural science is agronomy which deals with the study of crops production for food and fiber. It plays crucial role as it is the driving force of economy in most developing countries, feeding the humans directly and indirectly (Chhipa and Joshi 2016). There are predictions for world population to be nine billion in 2050. It will be really difficult to feed such a huge population with the same resources and deteriorated environment. Therefore, global production in the field of agriculture should increase to feed this rapidly increasing population. But agricultural sector has been facing serious challenges for sustainable food production (Godfray and Garnett 2014; Mcclung 2014).

Major problems faced by agriculture include increased population, climatic changes, soil erosion, and difference in soil conditions, micro and macro nutrient deficiencies, pathogens attack, urbanization and industrialization which affect the production of food. Almost 35–40% crop production depends upon the use of fertilizers. But excessive use of these synthetic fertilizers also affects the growth and yield of crops directly (Manjunatha et al. 2016). Disease management of crops with pesticides has led to the increase in concentration of toxic compounds in soil as well as ground and surface water. Conventional methods used for irrigation purposes are another reason of water depletion as more water is being pumped out than it is replenished (Rodell et al. 2009). Water scarcity all over the world due to variable climatic conditions and extreme weather actions has negative impact on crop

M. Ansari (✉) · S. Ahmed
Department of Botany, The University of Punjab, Lahore, Pakistan

K. Shahzadi
Department of Botany, Lahore College for Women University, Lahore, Pakistan

S. Javad (ed.), *Nanoagronomy*, https://doi.org/10.1007/978-3-030-41275-3_1

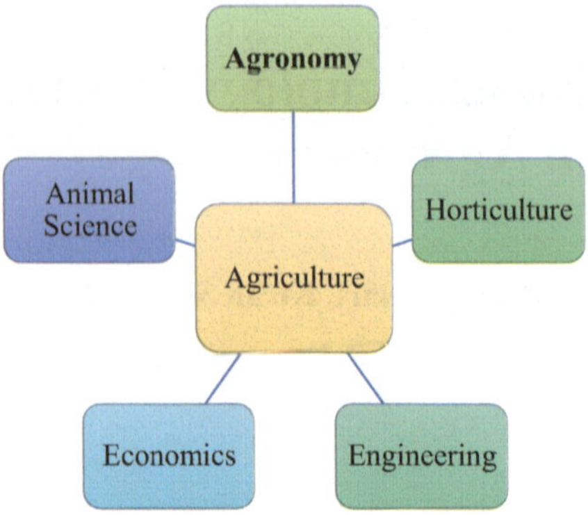

Fig. 1.1 Sub branches of agricultural science

production. Conventional irrigation methods also remove the important minerals from soil causing the salinity which ultimately leads to the reduction in agricultural land (Österholm and Åström 2004; Presley et al. 2004; Mukhopadhyay 2005). This scenario for such complex agricultural system is a serious challenge mostly for developing countries. Therefore, advancement in science and technology is needed.

Nanotechnology provides such tools and techniques which can revolutionize agricultural industry. Nanotechnology is the use of nanomaterials with exceptional properties to enhance the productivity of crops as well as livestock (Batsmanova et al. 2013). It is focused to

- improve the quality of food
- protect crop
- monitor the growth of plant
- enhance the production of food
- identify the disease-causing pathogens

Among above mentioned applications, food production and crop protection are the main applications of nanotechnology in agricultural industry. Innovative tools are provided by nanotechnology to deliver agrochemicals at targeted area safely without disturbing the ecosystem. It has developed such carrier systems that cause the control release of compounds when needed; that is how concentration of pesticides in the environment can be reduced (Ghormade et al. 2011; González et al. 2014).

1.2 What Is Nanotechnology?

The concept of nanotechnology was seeded at first by a physicist "Richard Phillips Feynman" in 1959 during his talk "*There's Plenty of Room at the Bottom*" in a conference held at the California Institute of Technology, Pasadena. He described the process theoretically for controlled manipulation of individual atoms and molecules. Professor Norio Taniguchi coined the term "nanotechnology" 10 years later at Tokyo State University. Nanotechnology has been modernized with the advancement of microscopes in 1981 as modern scanning microscopes so that individual atoms can be observed (Marchiol 2012). Nanotechnology is an advanced field of science, deals with production, manipulation and implication of matter at nanoscale. The name is based on a Greek letter “nano” meaning dwarf. It is one billionth part of a meter or 10^{-9} m (Holdren 2011; Rai and Ingle 2012).

Nanomaterials are very minute structures which range from 0.1 to 100 nm. These are very important because of these microscopic size as well as different properties from bulk material. Properties of these nanomaterials such as electrical conductance, magnetism, chemical reactivity, optical effects and physical strength vary from bulk materials due to their smaller size. These nanomaterials form a link between bulk material and their respective nanoparticles (Boisseau and Loubaton 2011).

1.3 Classification of Nanomaterials

Nanomaterials are classified intro three groups on dimensional basis (Hett 2004).

1.3.1 One-Dimensional Nanoparticles

Nanomaterials having less than 100 nm size with one dimension are grouped into this category. Nanowires and nanorods are examples of one-dimensional nanoparticles which are being used in buildup of various chemical and biological sensors, solar cells, IT systems and optical devices.

1.3.2 Two-Dimension Nanoparticles

Nanomaterials having size less than 100 nm along two dimensions at least are known as 2D nanoparticles e.g. carbon nanotubes fibers and platelets.

1.3.3 Three-Dimension Nanoparticles

Metallic nanomaterials having <100 nm in all dimensions i.e. quantum dots, dendrimers and hollow spheres are three dimensional nanoparticles.

Nanomaterials are also classified based on structural configuration.

- Metallic nanoparticles
- Nanocrystals Quantum dots

- Carbon Nanotubes
- Liposome
- Dendrimer
- Polymeric micelles
- Polymeric nanoparticles

1.4 Synthesis of Nanomaterials

Two approaches are being used to synthesize nanoparticles (Fig. 1.2) which are mentioned below:

1. Top to bottom approach
2. Bottom to up approach

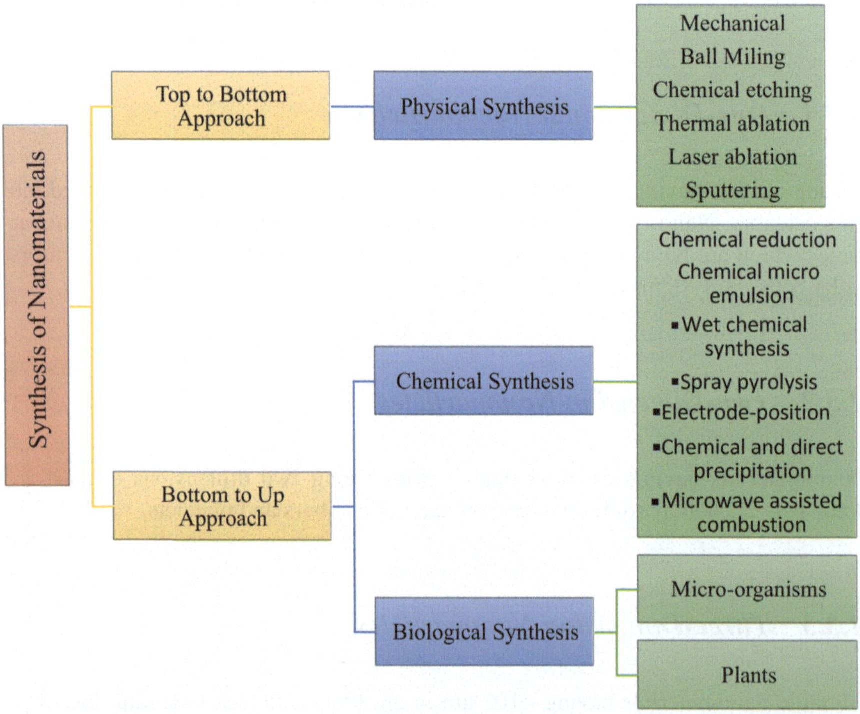

Fig. 1.2 Methods for synthesis of nanomaterials

1.4.1 Top to Bottom Approach

It refers to the breakdown of suitable given bulk material into particles in size range of nanometer due to reduction in size by various methods. This includes grinding of material, milling, sputtering laser ablation and evaporation–condensation. Various nanoparticles have been synthesized using this technique such as silver nanoparticles, gold nanoparticles, lead and fullerene nanoparticles. A tube furnace is used in this method to generate high atmospheric pressure. In the center of the tube furnace, concerned bulk material is placed on a boat, allowed to vaporize and carried through a gas. But use of this tube furnace has several disadvantages as it is a large equipment required huge place to be installed. A lot of energy more than several kilowatts is consumed to raise the temperature of tube furnace around the bulk material and it also entails a lot of time to attain a stable operating temperature and gain thermal stability. Imperfection of the surface structure of nanoparticles is the other limitation of this procedure as various physical properties are highly dependent on surface chemistry of nanoparticles (Samberg et al. 2009; Prathna et al. 2011; Sintubin et al. 2011; Kumar et al. 2014a).

1.4.2 Bottom to Up Approach

When nanoparticles are being synthesized using different chemicals as well as biological systems that is known as bottom to top approach. In bottom to top approach, atoms are self-assembled into new nuclei forming the particles of nano size.

1.4.2.1 Chemical Synthesis

Nanoparticles can be synthesized by a number of chemical methods as mentioned in Fig. 1.2 (Mitra et al. 2015; Yuvakkumar et al. 2015). But, the most commonly used method among all the chemical methods is chemical reduction for the synthesis of nanoparticles (Elghanian et al. 1997; Hurst et al. 2006). Various compounds including both organic and inorganic are being used as reducing agents for the production of nanoparticles i.e. sodium borohydride ($NaBH_4$), ascorbate, elemental hydrogen, Tollen's reagent, *N,N*-dimethyl formamide, sodium citrate, Tollen's reagent and copolymers of poly ethylene glycol (Tran and Le 2013: Iravani et al. 2014). Capping agents are the chemicals which are responsible to control and stabilize the size of the nanoparticles avoiding the aggregation. Nanoparticles can be synthesized in bulk amounts by using reduction capability of different chemicals and it takes very less time for reaction completion. But it becomes harmful due to use of synthetic chemicals which are toxic, hazardous and risk for environment and living bodies (Sastry et al. 2003). This reason leads to the development of nanoparticles by using methods other than chemical method. Therefore, need to develop such methods become mandatory which are non-toxic, environment friendly and economically beneficial.

1.4.2.2 Biological Synthesis

Biological synthesis of nanoparticles plays a vital role in field of nanotechnology. Use of biological entities as microorganisms including viruses, bacteria, fungi as well as plant material either in form of extract or biomass is an alternate way in an ecofriendly manner for production of nanoparticles than to the physical and chemical methods (Reddy et al. 2012).

Nanoparticle Synthesis Using Microorganisms

Microorganisms are important biological factories being used for synthesis of nanoparticles. Microorganisms produce several reductase enzymes which can reduce the metals into metallic nanoparticles having narrow range of size distribution. This approach for the synthesis of nanoparticles holds the immense potential as

1. It is ecofriendly, avoiding the use of toxic and harsh chemicals
2. It is a cost-effective tool which do not require consumption of high energy and longer time as physiochemical approaches

Microorganisms i.e. bacteria, yeasts and fungi have been considered for the synthesis of metal nanoparticles both extra and intracellularly (Table 1.1). Myco-synthesis is an easy and straightforward approach rather than the use of bacteria for biological synthesis of stable nanoparticles. Important metabolites are produced by most fungi having bioaccumulation ability that helps in synthesis of nanoparticles. Simple downstream processes are involved to culture fungi and are highly tolerant for metals. Metal salts have more binding capacity with fungal biomass than other microorganisms therefore help in high yield of nanoparticles (Castro-Longoria et al. 2011; Alghuthaymi et al. 2015). There are three mechanisms which are explained by scientists involve in the synthesis of nanoparticles with fungi are (Alghuthaymi et al. 2015):

1. Nitrate reductase action
2. Electron shuttle quinones
3. Both nitrate reductase action and electron shuttle quinones

Various fungal enzymes such as the enzymes of *Penicillium* species (reductase enzymes) and the enzymes of *Fusarium oxysporum* (nitrate reductase and -NADPH-dependent reductases) are reported to play a significant role in synthesis of nanoparticle (Kumar et al. 2007).

Nanoparticles synthesis by plant biomass and extracts is more beneficial than using microorganisms as keeping cultures of microbes involve complex procedures.

Table 1.1 List of microorganisms involved in nanoparticles' synthesis

Type of microorganisms	Name	Nanoparticle	Size (nm)	References
Fungi	*Alternaria alternata*	Silver	27–79	Baharvandi et al. (2014)
	Aspergillius species	Zinc	25	Raliya and Tarafdar (2014)
	Aspergillus niger	Silver	15–20	Kumar et al. (2008)
Bacteria	*Bacillus methylotrophicus*	Silver	10–30	Wang et al. (2016)
	Bhargavaea indica	Silver	111	Singh et al. (2015a, 2016a)
	Brevibacterium frigoritolerans	Silver	97	Singh et al. (2015b)
	Coriolus versicolor	Cadmium sulphide	20	Sanghi and Verma (2009)
	Fusarium oxysporum	Silver	5–15	Ahmad et al. (2003)
	Fusarium oxysporum	Zirconia	3–11	Bansal et al. (2004)
	Fusarium oxysporum f. sp. *lycopersici*	Platinum	10–50	Riddin et al. (2006)
	Ganoderma applanatum	Silver	20–25	Jogaiah et al. (2019)
	Hormoconis resinae	Gold	3–20	Mishra et al. (2010)
	Neurospora crassa	Cobalt	64	Rashmi et al. (2004)
	Penicillium fellutanum	Silver	5–25	Kathiresan et al. (2009)
	Pleurotus sajorcaju	Silver	5–50	Nithya and Ragunathan (2009)
	Pseudomonas deceptionensis	Silver	10–30	Jo et al. (2016)
	Trichoderma harzianum	Silver	15–25	Abboud (2018)
	Verticillium sp.	Magnetite	20–50	Bharde et al. (2006)
	Weissella oryzae	Silver	20–50	Singh et al. (2016b)

Green Synthesis

Nanoparticles synthesis using green method has developed as an advance technology now a day. It is very simple, an environment friendly and cost effective and easily scaled up for production of nanoparticles at large scale with the help of plants (Table 1.2). Consumption of energy to generate high pressure and temperature as well as toxic compounds are not needed in synthesis of nanoparticles by green method. Use of water as reducing medium makes it biocompatible (Noruzi 2015). Availability of plant materials in access which are nontoxic in nature also scaled up for production of nanoparticles at large scale and applicable in medical science, environmental areas and other biological fields. Various plant parts (leaves, fruits, seeds, stems and roots) including their extract rich in phytochemicals have been used for the synthesis of nanoparticles, which act as both reducing and capping agents (Ramesh et al. 2015; Xiao et al. 2016).

Table 1.2 List of plants involved in synthesis of NPs

Plant name		Nanoparticle	Size (nm)	Reference
Alternanthera dentate	Leaves	Silver	50–100	Kumar et al. (2014b)
Aloe vera	Leaves	Gold	–	Chandran et al. (2006)
Azadirachta indica	Leaves	Silver	10–35	Ramya and Subapriya (2012)
Coriandum sativum	Leaves	Gold	6.7–57.9	Narayanan and Sakthivel (2008)
Citrus sinensis	Peel	Silver	3–12	Veeraputhiran (2013)
Cuminum cyminum	Seed	Gold	1–10	Sneha et al. (2011)
Eucalyptus hybrid	Peel	Silver	50–150	Kaviya et al. (2011)
Euphorbia nivulia	Latex	Copper		Lee et al. (2013)
Ginko biloba	Leaves	Silver	15–500	Song and Kim (2009)
Medicago sativa (alfalfa)	Leaves	Titanium-nickel alloys	1–4	Schabes et al. (2006)
Medicago sativa (alfalfa)	Leaves	Iron oxide	2–10	Herrera et al. (2008)
Magnolia kobus	*Leaves*	Copper	37–110	Lee et al. (2013)
Ocimum sanctum	*Leaves*	Copper	77	Kulkarni and Kulkarni (2013)
Physalis alkekengi	Shoot	Zinc oxide	72.5	Qu et al. (2011b)
Sedum alfredii Hance	–	Zinc oxide	53.7	Qu et al. (2011a)
Syzygium aromaticum	Flower Buds	Copper	40–45	Subhankari and Nayak (2013a)
Zingiber officinale	Rhizome	Copper	40	Subhankari and Nayak (2013b)

1.5 Overview of Nano-technological Applications in Agriculture

The major reason for the application of nanotechnology in the field of agriculture is the production of food, feed and fiber. As the population is increasing tremendously, more food will be in demand while natural resources (land area, fertility of soil and water) are being depleted continuously. Therefore, increase in cost of production of crops at an alarming rate is expected due to limited resources. These limitations are needed to be overcome, and precision farming is much better solution to maximize the agricultural production and reduction of costs. Various advance techniques of nanotechnology are available which can be helpful in improving the practices of precision farming. Smart delivery of fertilizers and pesticides, detection of the contaminants and pathogens can be done by using suitable nanotools and sensors (Ahmed et al. 2013). Some of them are described as follows.

1.5.1 Crop Production and Growth

1.5.1.1 Seed Production

Seeds are meant to protect embryo "the mini plant". These are self-perpetuating biological entities able to survive in harsh environment. Progression in the development of new plant from a seed is very tedious especially in case of wind pollinated crops. Nanotechnology offers such tools which can be used to harness the full potential of seed. Detection of pollen load with help of nanosensors is a method that can ensure the genetic purity by avoiding contamination during its transfer from anther to stigma. The flight of pollen is determined by various factors i.e. air temperature, wind velocity, humidity and production of pollens.

Crop productivity can also be enhanced using genetically modified seeds as required genes are being incorporated into seeds and sold in the market with specific given codes. Nanobarcodes can also be used to track the sold seeds in markets as their codes are readable with machines (Nicewarner-Pena et al. 2001). Stored seeds are mostly killed by disease causing pathogens, but seed coatings with nanomaterials i.e. Ag (silver), Au (gold), Zn (zinc), Mn (manganese) and Ti (titanium) can also be used in far less quantities to protect them.

1.5.1.2 Seed Germination

Effects of different nanomaterials on seed germination and growth of plants have been studied in recent years. The main objective of this study is to promote the use of nanomaterials for agricultural practices. Penetration of nanomaterials into the seed is a key to the increased germination rate of seeds. Nanomaterials have abilities to increase the uptake of water and oxygen which fasten the germination rate and enhance the resistance against stress. Nutrients absorption, breakdown of the organic compounds and quenching of free radicals of oxygen, formed during the various processes is also accelerated by nanomaterials.

Another advancement done by nanotechnology is seed coating. Following advantages have been perceived in result of seed coating e.g.,

(a) It senses the availability of water and allow seeds to imbibe only when time is right for germination
(b) It detects the moisture content during storage to take proper measure to reduce the damage
(c) It determines the ageing of seeds

Scientists are working to improve the germination rate of seeds for rain fed crops by using carbon nanotubes and metallic nanoparticles. Improved germination of the tomato seeds by using carbon nanotube has been reported as carbon nanotube act as pores/channels allowing better penetration of water into the seeds. These processes ease the germination which can be exploited in rain fed agricultural system (Khodakovskaya et al. 2009).

1.5.1.3 Nanofertilizer

Growth and development of plants is dependent upon the nutrients which are not sufficiently available in soil. Therefore, fertilizers are being applied to fulfill the needs. But mostly these applied fertilizers are not readily available to plants, as decomposed in the soil, degraded or leached down during irrigation and water logging, and plants become nutrient deficient causing the yield lower than usual (Solanki et al. 2015). Conventional fertilizers can be surpassed by nanofertilizers, as in nanofertilizers nutrients are encapsulated/coated with thin film of nanomaterials or provided as emulsions or in form of nanoparticles (Derosa et al. 2010). Nanomaterials, especially metallic and carbon-based nanoparticles, are produced for improvement in growth and development of crops and better yield. These are also used for assimilation purpose, translocation of nutrients and storage of compounds (Nair et al. 2010). It has been reported that nanotechnology has positive impacts on morphological as well as physiological parameters of many crops including soybean (Agrawal and Rathore 2014), spinach (Gao et al. 2006) and peanut (Giraldo et al. 2014). Few impacts of them are given below:

- Improve germination percentage by increasing their water absorption ability
- Increase in root and shoot length
- Enhance the vegetative biomass
- Higher photosynthetic rate as then increase the light retention and light diffusion in the plant
- Nitrogen assimilation enhanced due to action of Nitrate reductase.

Water pollution can be reduced by use of nanofertilizers as nutrients are discharged in controlled way into the soil avoiding the leaching of nutrients and readily available to plants (Naderi and Abedi 2012). Utilization of nanofertilizers also reduces the soil toxicity and lessens the toxic effects of fertilizers taken in larger amount.

1.5.2 Crop Disease Management

Crop loss is the biggest issue worldwide mainly caused by insects (14%), weeds (13%) and various plant diseases (13%). The value of loss caused by plant diseases is estimated as 2000 billion dollars per year (Pimentel 2009). Use of synthetic chemicals is today's fastest, easiest and cheapest way to control loss causing agents. But many problems have been raised due to uncontrolled use of these chemicals including; (a) adverse effects on human health, (b) adverse effects on pollinating insects and domestic animals and (c) entering this material into the soil and water and its direct and indirect effects on ecosystems. Scientists have currently introduced methods of biological control but these proved as very expensive. Therefore, it is purposed that nanomaterials can be used to solve above mentioned problems (Sharon et al. 2010).

1.5.2.1 Nanopestiside

Various products; pesticides, insecticides and insect repellants can be prepared using nanoparticles as these are highly effective against pests as well as insects (Barik et al. 2008). Useful properties of nanomaterials i.e. stiffness, solubility, thermal stability, permeability, crystallinity and biodegradability are needed for formulating nanopesticides. They also will have larger surface area to volume ratio and greater absorption rate and higher affinity to the target (Bordes et al. 2009; Bouwmeester et al. 2009). Nanopesticides are either prepared in combination of nanoparticles with pesticidal active ingredients or nanoparticles with high pesticidal activity. Use of these nanopesticides can reduce the runoff of organic solvents and movement of unwanted pesticide by increasing the dispersion of these formulations (Bergeson 2010). Recently some of the techniques such as nanoencapsulates, nanoemulsions, nanocontainers and nanocages have been reported for delivery of nanopesticide to protect plants from pest attack (Bouwmeester et al. 2009; Lyons and Scrinis 2009; Bergeson 2010).

Variety of products is used as nanopesticides including (Matthews 2000):

- Nanoemulsions
- Metallic oxide nanoparticles such as titanium dioxide
- Organic ingredients active ingredient
- Polymer-based inorganic silica-based nanoparticles
- Nanoclays in various forms

1.5.2.2 Nanofungicide

Fungi are an important group of organisms that include plant pathogens causing approximately $45 billion loss in crops all over the world every year. These pathogens not only affect the seedling emergence by attacking the seeds but also able to attack the plant tissues at different growth stages and lowering the yield ultimately (Fernandez-Acero et al. 2007). Conventional methods are being used for decades to control these plant pathogens, but excessive use of these methods has affected both the economy and environment. Applied fungicides can be lost (90%) as result of overflow and enter in the ecosystem disturbing both the agriculture and environment during its application in the open field. Therefore, farmers need to apply these fungicides in huge amount, and it becomes costly for them. Overuse of fungicide causes environmental problems which is a matter of concern for scientists as well as public. Each year almost 2.5 million tons of fungicides are used to protect crops. Universal harm of these fungicides has reached $100 billion annually (Pimentel 2009). There are two main reason of harm given below (Koul et al. 2008):

1. These are highly toxic, affect the health of both humans and animals
2. Non-biodegradable therefore persists in environment contaminating the soil and water resources

By considering the above mentioned factors, development of environment friendly fungicide like bio-based materials or nanomaterials is needed urgently. Use of nanomaterials can solve the problems as nanofungicides can be used against these fungal pathogens. Nanofungicide is a term used for the any fungicide entities ranges in size of 1–100 nm. These nanofungicides have novel properties due to their extreme small size. Use of these nanofungicides in crop protection should be evaluated first and health of public should be the matter of concern. Nanosensors and devices can be used to monitor the intelligent delivery of fungicide on targeted site (Matthews 2000).

1.5.2.3 Nanoherbicide

Weeds are unwanted plants growing in soil and taking the essential nutrients which are added for the growth and development of crops to get better yield. These weeds not only survive but spread in the soil through tubers and deep roots (underground features). Conventional methods are available to remove weeds as removal by hands, but these are laborious and time-consuming jobs. Seed banks of weeds in the soil can be destroyed and their germination can be prevented while the conditions become suitable for their growth. For this purpose, nanoherbicides can be used as being very small they will blend with soil easily and eradicate weeds in an eco-friendly way without leaving any toxic residues and prevent the growth of weeds. Herbicides can be applied in the form of an active ingredient combined with any smart delivery system purposed by nanotechnology according to requirement. Use of nanoherbicides can improve the crop production by reducing the competition for nutrients between the weeds and plants without harming the agricultural land, environment and the workers who have direct exposure to weeds while working in the field (Mukherjee et al. 2015).

1.5.2.4 Nanofibers

Nanotechnology also plays its role in recycling of residual materials from processing of agricultural products. Nanotechnology is used to produce valuable materials from agricultural processing waste. These products are of commercial importance and very useful to the industry. One of the examples is production of cellulose fibers during cotton processing which is also useful to produce low value products like cotton balls and yarn. Nanotechnology helps us to produce Nanofibers (100 nm in diameter) by electrospinning which can be used as nanofertilizers or pesticide adsorbent. These adsorbents are highly efficient due to their increased surface area and are really active to release the desired compounds at targeted location and time (Lang et al. 2003).

Nanofibers have another application of covering the chemical pesticides which prevent their scattering into the environment. This lowers the hazardous effects of chemical/pesticides. Such wise use of nanotechnology makes the use of pesticides more durable and secure. When some hydrophobic pollutants are added into water or soil by any source, they can be adsorbed by nanofibers. Nanofibers are also considered as the excellent source of separating organic pollutants from soil and water. Nanofibers based fabric is also being made for cleansing which has embedded antibodies. Such nano-fabric acts as detection tool to capture and isolate pathogens from surface. Fabric when comes in contact with any surface, is then tested for presence of pathogens. It has been also investigated for color change response on reaction with pathogens (Hager 2011). Its practical applications in near future can be revolutionary.

1.5.3 Nanoparticles in Post-harvest Disease Management

Human population is increasing many folds day by day causing depletion of natural resources. Supply of healthy food for that much population has become a daunting task. Efficiency in food production and decrease in wastage of post-harvest products due to resistant pathogens with the application of advanced technology such as biotechnology and nanotechnology is really needed. Various applications of nanotechnology in field of agriculture, horticulture and management of postharvest diseases have been described (Yadollahi et al. 2009; Chowdappa and Gowda 2013) as packing with nanomaterials:

- Increases the shelf life of products
- Controls the growth of pathogens by nanofilms and nano-coatings
- Limits the harmful effects of gases and rays i.e. UV-rays
- Detect the quality of food and its spoilage rate

Tools and techniques of nanotechnology can be used to dry, store and preservation of the products. Chitosan is a product of nanotechnology "Derivative of Chitin" is used in reducing the decay of fruits and vegetables as it is very effective in controlling the growth of most phytopathogenic fungi causing spoilage of fruit and vegetables after harvest (Hirano 1997; Liu et al. 2007).

1.5.4 Nanoscale Carriers

Delivery of various agricultural products such as pesticides, fungicides, herbicides, fertilizers and growth regulators hormones can be done by using nanoscale carriers. The mechanisms which are involved in the efficient delivery of products, controlled release and better storage includes

- Encapsulation and entrapment
- Polymers and dendrimers
- Surface ionic and weak bond attachments

Nanoscale carriers are not only involved in stability of products in the environment but also reduce the degradation and chemical runoff. These all factors result in reduction of cost of the product and environmental problems as well (Johnston 2010).

1.5.5 Nanosensors

Nano-technological advancements have familiarized the agricultural world with nanosensors which can be used to check the crops in fields. Nanosensors can be helpful in precision farming by increasing the productivity of agricultural crops with better management of time, fertilization and environment (Scott and Chen 2013). Use of nanosensors and devices in field of agriculture are given below:

1. The environmental factors such as weather conditions (temperature, humidity, air velocity and air quality) can be measured accurately
2. Level of water in the field and time for its requirement can be examined
3. Helpful in efficient use of the applied fertilizers, pesticides, herbicides and other treatments can also be monitored
4. Fertility of soil and its suitability for the crops can be investigated
5. Nutrient status of crop can be analyzed
6. Presence of any pathogen like pests, insects or microorganisms can be detected by nanodevices dispersed in the field
7. Evidence of stresses such as salt stress or drought can be collected
8. Optimal time for the plantation of crops and its harvesting can be determined

Nanosensors are used for real time monitoring for precision farming by collecting the essential data from the cultivated fields. Wireless nanosensors are linked with satellite systems for data collection i.e. geographic information and remote sensing devices can be dispersed in the field (Brock et al. 2011).

1.5.6 Sustainable Water Use

Scientists have been working on optimized use of water in agriculture and its conservation for future. Various approaches have been introduced i.e. sprinkler irrigation, gun and drip irrigation. More precise systems are needed to be developed for delivery of water in the field. Water holding ability of the soil should be increased to prevent the leaching of water. Proper distribution of water near the roots is a key factor of maximum water absorption by plants (Cross et al. 2009). Nano-hydrogels, a product introduced by nanotechnology, can be used for efficient use of water. These gels can absorb more water than normal soil and release it on demand. These can be used to store rain and irrigation water. It is especially useful in dry areas. This is highly needed as drought is considered the largest environmental risk for crop production. Nanosensors, distributed in the field, can measure the amount of already present water and determine the time of its requirement (Vundavalli et al. 2015).

1.5.7 Removal of Heavy Metals

Industrial growth has developed several environmental problems due to discharge of its waste including heavy metals and other pollutants directly into the soil and water. Heavy metals such as arsenic, cadmium, chromium, copper, lead, mercury, nickel, selenium, silver, zinc etc. are accumulating to the soil continuously. Removal of heavy metals takes place by leaching, erosion and plant uptake. But they cannot be broken down into other non-toxic or less toxic forms and eventually they survive in ecosystem. Their presence is major concern for public health as they are not only cytotoxic at very low concentration but mutagenic in nature thus lead to the cancers (Dixit et al. 2015).

Land and water bodies affected by heavy metals need to be rectified in order to make them contamination free. A number of techniques are already in use. But these techniques are only effective when concentration of contaminants is higher than 100 mg/L (Ahluwalia and Goyal 2007). These techniques include reverse osmosis, redox reactions, chemical precipitations, membrane filtration, electrochemical treatment and evaporation etc. But now nanotechnology provides a tool to identify and eliminate heavy metals from environment when their concentration is below 100 mg/L. Nanocoating with ligands is an important technique for this purpose. Ligands have higher absorption affinity for heavy metals. It may also prove to be a cost effective technique (Farmen 2009).

1.5.8 Hydroponics

Hydroponics is a technique of agriculture which is not very well known for common man. But it is also a fact that a number of fruits and vegetables in super stores are products of hydroponics. It is a branch of agriculture including the technique of growing plants without soil. It's a widely used technique. It is being used for production of lettuce, tomatoes, cucumber, melons, broccoli, sweet pepper, chilies and eggplant etc. Research is also going on hydroponic procedure of biofuel and fodder crops. Scientists have used nanotechnology to harvest nanoparticles in crops as nutrients. It is the future of man when agricultural land is going to be limited (Giordani et al. 2012; Schwabe et al. 2013; Sekhon 2014).

Nutrient management in agricultural production is increasingly important and is more effective in hydroponic than in soil-based production (Sekhon 2014). A recent work on nanophosphor-based on electroluminescence lighting device has shown that its use can reduce energy consumption significantly. Such nanotechnology-based light could reduce energy costs and encourage photosynthesis in indoor, hydroponic agriculture (Witanachchi et al. 2012).

1.5.9 Nanoparticles in Plant Tissue Culture

Nanotechnology for tissue engineering applications is just started, and most of the applications are still in the idea stage. To date, the applications are limited to the following: design of biomaterial scaffolds to provide controlled cell adhesion, modulating the spatial organization of cells in the scaffold, regulating the biomolecule in the local environment, self-assembling scaffolds, using a "lab-on-a-chip" for a better understanding the mechanisms of cell differentiation and propagation and tissue development and formation; combination of biomolecule functionality with synthetic or natural biomaterial scaffold for better regeneration; surface modification at nanoscale level for enhanced biological compatibility and activity; manipulation and detection of a single molecule; controlled and targeted drug delivery, the fabrication of highly organized scaffold, and cellular and genetic engineering of functional cell types. Although there are a lot of unknowns, nano-tissue engineering is not just a dream but quickly becoming reality. There is plenty space for the practical applications of nanotechnology in tissue engineering. Tissue engineering will be a perfect field for exploring the practical applications of nanotechnology.

1.6 Conclusion

It is concluded that nanotechnology has a great breakthrough in the field of agricultural sciences/agronomy as nanomaterials can be applied to fasten the germination/production of plants. Plant protection is another environment friendly application of effective nanoparticles in comparison to traditional approaches. Nanoscale carriers can deliver the agricultural products (pesticides and fertilizers) to the targeted sites without their degradation and leaching by chemical runoff. Nanosensors and devices can be used for real time monitoring for precision farming by collecting the essential data from the cultivated fields. These are also able to detect the presence of disease-causing pathogens in the field. Further research is needed to be done for practical implementations of this technology.

References

Abboud AMA (2018) Fungal biosynthesis of silver nanoparticles and their role in control of *Fusarium* wilt of sweet pepper and soil-borne fungi *in vitro*. Int J Pharmacol 14(6):773–780

Agrawal S, Rathore P (2014) Nanotechnology pros and cons to agriculture: a review. Int J Curr Microbiol App Sci 3(3):43–55

Ahluwalia SS, Goyal D (2007) Microbial and plant derived biomass for removal of heavy metals from wastewater. Bioresour Technol 98(12):2243–2257

Ahmad A, Mukherjee P, Senapati S, Mandal D, Khan MI, Kumar R, Sastry M (2003) Extracellular biosynthesis of silver nanoparticles using the fungus *Fusarium oxysporum*. Colloids Surf B: Biointerfaces 28(4):313–318

Ahmed F, Arshi N, Kumar S (2013) Chapter 11: Nanobiotechnology: scope and potential crop improvement. In: Crop improvement under adverse conditions. Springer, New York, NY, pp 245–269

Alghuthaymi MA, Almoammar H, Rai M, Said-Galiev E, Abd-Elsalam KA (2015) Myconanoparticles: synthesis and their role in phytopathogens management. Biotechnol Biotechnol Equip 29(2):221–236

Baharvandi A, Soleimani MJ, Zamani P (2014) Mycosynthesis of nanosilver particles using extract of *Alternaria alternate*. Arch Phytopathol Plant Protect 48(4):313–318

Bansal V, Rautaray D, Ahmad A, Sastry M (2004) Biosynthesis of zirconia nanoparticles using the fungus *Fusarium oxysporum*. J Mater Chem 14(22):3303–3305

Barik TK, Sahu B, Swain V (2008) Nanosilica—from medicine to pest control. Parasitol Res 103(2):253–258

Batsmanova LM, Gonchar LM, Taran NY, Okanenko AA (2013) Using a colloidal solution of metal nanoparticles as micronutrient fertilizer for cereals. Doctoral dissertation, Sumy State University

Bergeson LL (2010) Nanosilver: US EPA's pesticide office considers how best to proceed. Environ Qual Manag 19(3):79–85

Bharde A, Rautaray D, Bansal V, Ahmad A, Sarkar I, Yusuf SM et al (2006) Extracellular biosynthesis of magnetite using fungi. Small 2(1):135–141

Boisseau P, Loubaton B (2011) Nanomedicine, nanotechnology in medicine. C R Phys 12(7):620–636

Bordes P, Pollet E, Avérous L (2009) Nano-biocomposites: biodegradable polyester/nanoclay systems. Prog Polym Sci 34(2):125–155

Bouwmeester H, Dekkers S, Noordam MY, Hagens WI, Bulder AS, De Heer C et al (2009) Review of health safety aspects of nanotechnologies in food production. Reg Toxicol Pharm 53(1):52–62

Brock DA, Douglas TE, Queller DC, Strassmann JE (2011) Primitive agriculture in a social amoeba. Nature 469(7330):393–396

Castro-Longoria E, Vilchis-Nestor AR, Avalos-Borja M (2011) Biosynthesis of silver, gold and bimetallic nanoparticles using the filamentous fungus *Neurospora crassa*. Colloids Surf B: Biointerfaces 83(1):42–48

Chandran SP, Chaudhary M, Pasricha R, Ahmad A, Sastry M (2006) Synthesis of gold nanotriangles and silver nanoparticles using *Aloe vera* plant extract. Biotechnol Prog 22(2):577–583

Chhipa H, Joshi P (2016) Nanofertilizers, nanopesticides and nanosensors in agriculture. In: Nanoscience in food and agriculture, vol 1. Springer, Cham, pp 247–282

Chowdappa P, Gowda S (2013) Nanotechnology in crop protection: status and scope. Pest Manag Hortic Ecosyst 19(2):131–151

Cross KM, Lu Y, Zheng T, Zhan J, McPherson G, John V (2009) Water decontamination using iron and iron oxide nanoparticles. In: Nanotechnology applications for clean water. William Andrew Publishing, New York, pp 347–364

DeRosa MC, Monreal C, Schnitzer M, Walsh R, Sultan Y (2010) Nanotechnology in fertilizers. Nat Nanotechnol 5(2):91

Dixit R, Malaviya D, Pandiyan K, Singh U, Sahu A, Shukla R et al (2015) Bioremediation of heavy metals from soil and aquatic environment: an overview of principles and criteria of fundamental processes. Sustainability 7(2):2189–2212

Elghanian R, Storhoff JJ, Mucic RC, Letsinger RL, Mirkin CA (1997) Selective colorimetric detection of polynucleotides based on the distance-dependent optical properties of gold nanoparticles. Science 277(5329):1078–1081

Farmen L (2009) Commercialization of nanotechnology for removal of heavy metals in drinking water. In: Nanotechnology applications for clean water. William Andrew Publishing, New York, pp 115–130

Fernandez-Acero FJ, Carbú M, Garrido C, Vallejo I, Cantoral JM (2007) Proteomic advances in phytopathogenic fungi. Curr Proteom 4(2):79–88

Gao F, Hong F, Liu C, Zheng L, Su M, Wu X et al (2006) Mechanism of nano-anatase TiO_2 on promoting photosynthetic carbon reaction of spinach. Biol Trace Element Res 111(1–3):239–253

Ghormade V, Deshpande MV, Paknikar KM (2011) Perspectives for nano-biotechnology enabled protection and nutrition of plants. Biotechnol Adv 29(6):792–803

Giordani T, Fabrizi A, Guidi L, Natali L, Giunti G, Ravasi F et al (2012) Response of tomato plants exposed to treatment with nanoparticles. Int J Environ Qual 8(8):27–38

Giraldo JP, Landry MP, Faltermeier SM, McNicholas TP, Iverson NM, Boghossian AA et al (2014) Plant nanobionics approach to augment photosynthesis and biochemical sensing. Nat Mater 13(4):400–408

Godfray HCJ, Garnett T (2014) Food security and sustainable intensification. Philos Trans R Soc B Biol Sci 369(1639):20120273

González JOW, Gutiérrez MM, Ferrero AA, Band BF (2014) Essential oils nanoformulations for stored-product pest control–characterization and biological properties. Chemosphere 100:130–138

Hager H (2011). Nanotechnology in agriculture. http://www.topcropmanager.com

Herrera BR, Zorrilla C, Rius JL, Ascencio JA (2008) Electron microscopy characterization of biosynthesized iron oxide nanoparticles. Appl Phys A 91(2):241–246

Hett A (2004) Nanotechnology: small matter, many unknowns. Swiss Reinsurance Company, Zurich

Hirano S (1997) Application of chitin and chitosan in the ecological and environmental fields. In: MFA G (ed) Application of chitin and chitosan. CRC Book, Boca Raton, FL

Holdren JP (2011) The national nanotechnology initiative strategic plan report at subcommittee on nanoscale science, engineering and technology of committee on technology. National Science Technology Council (NSTC), Arlington

Hurst SJ, Lytton-Jean AK, Mirkin CA (2006) Maximizing DNA loading on a range of gold nanoparticle sizes. Anal Chem 78(24):8313–8318

Iravani S, Korbekandi H, Mirmohammadi SV, Zolfaghari B (2014) Synthesis of silver nanoparticles: chemical, physical and biological methods. Res Pharm Sci 9(6):385–406

Jo JH, Singh P, Kim YJ, Wang C, Mathiyalagan R, Jin CG, Yang DC (2016) *Pseudomonas deceptionensis* DC5-mediated synthesis of extracellular silver nanoparticles. Artif Cells Nanomed Biotechnol 44(6):1576–1581

Jogaiah S, Kurjogi M, Abdelrahman M, Hanumanthappa N, Tran LSP (2019) *Ganoderma applanatum*-mediated green synthesis of silver nanoparticles: structural characterization, and *in vitro* and *in vivo* biomedical and agrochemical properties. Arab J Chem 12(7):1108–1120

Johnston CT (2010) Probing the nanoscale architecture of clay minerals. Clay Miner 45(3):245–279

Kathiresan K, Manivannan S, Nabeel MA, Dhivya B (2009) Studies on silver nanoparticles synthesized by a marine fungus, *Penicillium fellutanum* isolated from coastal mangrove sediment. Colloids Surf B: Biointerfaces 71(1):133–137

Kaviya S, Santhanalakshmi J, Viswanathan B, Muthumary J, Srinivasan K (2011) Biosynthesis of silver nanoparticles using *Citrus sinensis* peel extract and its antibacterial activity. Spectrochim Acta A Mol Biomol Spectrosc 79(3):594–598

Khodakovskaya M, Dervishi E, Mahmood M, Xu Y, Li Z, Watanabe F, Biris AS (2009) Carbon nanotubes are able to penetrate plant seed coat and dramatically affect seed germination and plant growth. ACS Nano 3(10):3221–3227

Koul O, Walia S, Dhaliwal GS (2008) Essential oils as green pesticides: potential and constraints. Biopestic Int 4(1):63–84

Kulkarni VD, Kulkarni PS (2013) Green synthesis of copper nanoparticles using *Ocimum sanctum* leaf extract. Int J Chem Stud 1(3):1–4

Kumar SA, Abyaneh MK, Gosavi SW, Kulkarni SK, Pasricha R, Ahmad A, Khan MI (2007) Nitrate reductase-mediated synthesis of silver nanoparticles from $AgNO_3$. Biotechnol Lett 29(3):439–445

Kumar R, Liu D, Zhang L (2008) Advances in proteinous biomaterials. J Biobased Mat Bioenergy 2(1):1–24

Kumar DA, Palanichamy V, Roopan SM (2014a) Green synthesis of silver nanoparticles using *Alternanthera dentata* leaf extract at room temperature and their antimicrobial activity. Spectrochim Acta A Mol Biomol Spectrosc 127:168–171

Kumar PV, Pammi SVN, Kollu P, Satyanarayana KVV, Shameem U (2014b) Green synthesis and characterization of silver nanoparticles using *Boerhaavia diffusa* plant extract and their anti bacterial activity. Indus Crops Prod 52:562–566

Lang H, May RA, Iversen BL, Chandler BD (2003) Dendrimer-encapsulated nanoparticle precursors to supported platinum catalysts. J Am Chem Soc 125(48):14832–14836

Lee HJ, Song JY, Kim BS (2013) Biological synthesis of copper nanoparticles using *Magnolia kobus* leaf extract and their antibacterial activity. J Chem Technol Biotechnol 88(11):1971–1977

Liu J, Tian S, Meng X, Xu Y (2007) Effects of chitosan on control of postharvest diseases and physiological responses of tomato fruit. Postharvest Biol Technol 44(3):300–306

Lyons K, Scrinis G (2009) Under the regulatory radar? Nanotechnologies and their impacts for rural Australia. Tracking rural change: community, policy and technology in Australia, New Zealand and Europe. Australian National University E Press, Canberra, pp 151–171

Manjunatha SB, Biradar DP, Aladakatti YR (2016) Nanotechnology and its applications in agriculture: a review. J Farm Sci 29(1):1–13

Marchiol L (2012) Synthesis of metal nanoparticles in living plants. Ital J Agron 7:e37

Matthews GA (2000) Pests, pesticides and pest management. In: Mason J (ed) Highlights in environmental research. Imperial College Press, London, pp 165–189

McClung CR (2014) Making hunger yield. Science 344(6185):699–700

Mishra AN, Bhadauria S, Gaur MS, Pasricha R (2010) Extracellular microbial synthesis of gold nanoparticles using fungus *Hormoconis resinae*. JOM 62(11):45–48

Mitra S, Patra P, Pradhan S, Debnath N, Dey KK, Sarkar S et al (2015) Microwave synthesis of ZnO@ $mSiO_2$ for detailed antifungal mode of action study: understanding the insights into oxidative stress. J Coll Interf Sci 444:97–108

Mukherjee A, Sinha I, Das R (2015) Application of nanotechnology in agriculture: future prospects. In: Outstanding Young Chemical Engineers (OYCE) Conference, March, pp. 13–14

Mukhopadhyay SS (2005) Weathering of soil minerals and distribution of elements: pedochemical aspects 1. Clay Res 24(2):183–199

Naderi MR, Abedi A (2012) Application of nanotechnology in agriculture and refinement of environmental pollutants. J Nanotechnol 11(1):18–26

Nair R, Varghese SH, Nair BG, Maekawa T, Yoshida Y, Kumar DS (2010) Nanoparticulate material delivery to plants. Plant Sci 179(3):154–163

Narayanan KB, Sakthivel N (2008) Coriander leaf mediated biosynthesis of gold nanoparticles. Mater Lett 62(30):4588–4590

Nicewarner-Pena SR, Freeman RG, Reiss BD, He L, Peña DJ, Walton ID et al (2001) Submicrometer metallic barcodes. Science 294(5540):137–141

Nithya R, Ragunathan R (2009) Synthesis of silver nanoparticle using *Pleurotus sajor* caju and its antimicrobial study. Digest J Nanomater Biostruct 4(4):623–629

Noruzi M (2015) Biosynthesis of gold nanoparticles using plant extracts. Bioprocess Biosyst Eng 38(1):1–14

Österholm P, Åström M (2004) Quantification of current and future leaching of sulfur and metals from boreal acid sulfate soils, western Finland. Soil Res 42(6):547–551

Pimentel D (2009) Pesticides and pest control. In: Integrated pest management: innovation-development process. Springer, Dordrecht, pp 83–87

Prathna TC, Chandrasekaran N, Raichur AM, Mukherjee A (2011) Kinetic evolution studies of silver nanoparticles in a bio-based green synthesis process. Colloids Surf A Physicochem Eng Asp 377(1–3):212–216

Presley DR, Ransom MD, Kluitenberg GJ, Finnell PR (2004) Effects of thirty years of irrigation on the genesis and morphology of two semiarid soils in Kansas. Soil Sci Soc Am J 68(6):1916–1926

Qu J, Luo C, Hou J (2011a) Synthesis of ZnO nanoparticles from Zn-hyperaccumulator (*Sedum alfredii* Hance) plants. Micro Nano Lett 6(3):174–176

Qu J, Yuan X, Wang X, Shao P (2011b) Zinc accumulation and synthesis of ZnO nanoparticles using *Physalis alkekengi* L. Environ Pollut 159(7):1783–1788

Rai M, Ingle A (2012) Role of nanotechnology in agriculture with special reference to management of insect pests. Appl Microbiol Biotechnol 94(2):287–293

Raliya R, Tarafdar JC (2014) Biosynthesis and characterization of zinc, magnesium and titanium nanoparticles: an eco-friendly approach. Int Nano Lett 4(1):93

Ramesh M, Anbuvannan M, Viruthagiri G (2015) Green synthesis of ZnO nanoparticles using *Solanum nigrum* leaf extract and their antibacterial activity. Spectrochim Acta A Mol Biomol Spectrosc 136:864–870

Ramya M, Subapriya MS (2012) Green synthesis of silver nanoparticles. Int J Pharm Med Biol Sci 1(1):54–61

Rashmi K, Krishnaveni T, Ramanamurthy S, Mohan PM (2004) Characterization of cobalt nanoparticle from a cobalt resistant strain of *Neurospora crassa*. In: International symposium of research students on materials science and engineering, Dec 2004, Chennai

Reddy GAK, Joy JM, Mitra T, Shabnam S, Shilpa T (2012) Nano silver–a review. Int J Adv Pharm 2(1):09–15

Riddin TL, Gericke M, Whiteley CG (2006) Analysis of the inter- and extracellular formation of platinum nanoparticles by *Fusarium oxysporum* f. sp. *lycopersici* using response surface methodology. Nanotechnology 17(14):3482

Rodell M, Velicogna I, Famiglietti JS (2009) Satellite-based estimates of groundwater depletion in India. Nature 460(7258):999–1002

Samberg ME, Oldenburg SJ, Monteiro-Riviere NA (2009) Evaluation of silver nanoparticle toxicity in skin *in vivo* and keratinocytes *in vitro*. Environ Health Perspect 118(3):407–413

Sanghi R, Verma P (2009) A facile green extracellular biosynthesis of CdS nanoparticles by immobilized fungus. Chem Eng J 155(3):886–891

Sastry M, Ahmad A, Khan MI, Kumar R (2003) Biosynthesis of metal nanoparticles using fungi and actinomycete. Curr Sci 85(2):162–170

Schabes PS, Canizal G, Herrera-Becerra R, Zorrilla C, Liu HB, Ascencio JA (2006) Biosynthesis and characterization of Ti/Ni bimetallic nanoparticles. Opt Mater 29(1):95–99

Schwabe F, Schulin R, Limbach LK, Stark W, Bürge D, Nowack B (2013) Influence of two types of organic matter on interaction of CeO_2 nanoparticles with plants in hydroponic culture. Chemosphere 91(4):512–520

Scott N, Chen H (2013) Nanoscale science and engineering for agriculture and food systems. Ind Biotechnol 9(1):17–18

Sekhon BS (2014) Nanotechnology in agri-food production: an overview. Nanotechnol Sci Appl 7:31

Sharon M, Choudhary AK, Kumar R (2010) Nanotechnology in agricultural diseases and food safety. J Phytol 4:83–92

Singh P, Kim YJ, Singh H, Wang C, Hwang KH, Farh MEA, Yang DC (2015a) Biosynthesis, characterization, and antimicrobial applications of silver nanoparticles. Int J Nanomedicine 10:2567

Singh P, Kim YJ, Singh H, Mathiyalagan R, Wang C, Yang DC (2015b) Biosynthesis of anisotropic silver nanoparticles by *Bhargavaea indica* and their synergistic effect with antibiotics against pathogenic microorganisms. J Nanomater 4:234741

Singh P, Kim YJ, Wang C, Mathiyalagan R, Yang DC (2016a) Microbial synthesis of flower-shaped gold nanoparticles. Artif Cells Nanomed Biotechnol 44(6):1469–1474

Singh P, Kim YJ, Wang C, Mathiyalagan R, Yang DC (2016b) *Weissella oryzae* DC6-facilitated green synthesis of silver nanoparticles and their antimicrobial potential. Artif Cells Nanomed Biotechnol 44(6):1569–1575

Sintubin L, De Gusseme B, Van der Meeren P, Pycke BF, Verstraete W, Boon N (2011) The antibacterial activity of biogenic silver and its mode of action. Appl Microbiol Biotechnol 91(1):153–162

Sneha K, Sathishkumar M, Lee SY, Bae MA, Yun YS (2011) Biosynthesis of Au nanoparticles using cumin seed powder extract. J Nanosci Nanotechnol 11(2):1811–1814

Solanki P, Bhargava A, Chhipa H, Jain N, Panwar J (2015) Nano-fertilizers and their smart delivery system. In: Nanotechnologies in food and agriculture. Springer, Cham, pp 81–101

Song JY, Kim BS (2009) Rapid biological synthesis of silver nanoparticles using plant leaf extracts. Bioprocess Biosyst Eng 32(1):79–84

Subhankari I, Nayak PL (2013a) Synthesis of copper nanoparticles using *Syzygium aromaticum* (cloves) aqueous extract by using green chemistry. World J Nano Sci Technol 2(1):14–17

Subhankari I, Nayak PL (2013b) Antimicrobial activity of copper nanoparticles synthesized by ginger (*Zingiber officinale*) extract. World J Nano Sci Technol 2(1):10–13

Tran QH, Le AT (2013) Silver nanoparticles: synthesis, properties, toxicology, applications and perspectives. Adv Nat Sci Nanosci Nanotechnol 4(3):033001

Veeraputhiran V (2013) Bio-catalytic synthesis of silver nanoparticles. Int J Chem Tech Res 5(5):2555–2562

Vundavalli R, Vundavalli S, Nakka M, Rao DS (2015) Biodegradable nano-hydrogels in agricultural farming-alternative source for water resources. Procedia Mater Sci 10:548–554

Wang C, Kim YJ, Singh P, Mathiyalagan R, Jin Y, Yang DC (2016) Green synthesis of silver nanoparticles by *Bacillus methylotrophicus* and their antimicrobial activity. Artif Cells Nanomed Biotechnol 44(4):1127–1132

Witanachchi S, Merlak M, Mahawela P (2012) Nanotechnology solutions to greenhouse and urban agriculture. Technol Innov 14(2):209–217

Xiao L, Liu C, Chen X, Yang Z (2016) Zinc oxide nanoparticles induce renal toxicity through reactive oxygen species. Food Chem Toxicol 90:76–83

Yadollahi A, Arzani K, Khoshghalb H (2009) The role of nanotechnology in horticultural crops postharvest management. In: Southeast Asia symposium on quality and safety of fresh and fresh-cut produce, vol 875, pp 49–56

Yuvakkumar R, Suresh J, Saravanakumar B, Nathanael AJ, Hong SI, Rajendran V (2015) Rambutan peels promoted biomimetic synthesis of bioinspired zinc oxide nanochains for biomedical applications. Spectrochim Acta A Mol Biomol Spectrosc 137:250–258

Chapter 2
Nanotechnology and Plant Tissue Culture

Amina Tariq, Saiqa Ilyas, and Shagufta Naz

2.1 Introduction

Nanotechnology is the study of managing minute particles to make new materials or modifying existing ones (Bhushan and Kumar 2010). Professor Norio Taniguchi at the University of Tokyo in 1974 first designed the expression "nanotechnology". Nanotechnology was defined as "an innovation to obtain the additional high presicion and ultra fine measurements, for example the exactness and fineness in the range of 1 nanometer (nm or 10^{-9} m long)".

The term Nano of nanomaterials was originated from the Greek word signifying "dwarf". More absolutely, the word nano implies 10^{-9} or one billionth of a meter (Huang et al. 2015) and is utilized for materials with a size extending somewhere in the range of 1–100 nm (Rai and Ingle 2012). Nanomaterials are usually having very specific and special characters owing to their very small size which also increases their surface area, which in turn provides more interactive surface. These tiny sized nanoparticles have a lot of size dependent characteristics which are not exhibited by their macro-sizes. Optical properties and absorbing capacity are some of the examples which may add to their increased activity as antibiotic, fertilizer or pesticide (Aslani et al. 2014).

Nanoparticles can be separated into two main groups on the basis of their origin i.e., natural and anthropogenic. Natural nanoparticles are part of structure of living bodies and are made by the nature in the living body while anthropogenic are the produced or designed nanoparticles. These anthropogenic nanoparticles can be further subdivided into two main groups as organic nanoparticles (carbon-containing) and inorganic nanoparticles. The above said anthropogenic type of nanoparticles

A. Tariq (✉)
Department of Botany, Lahore College for Women University, Lahore, Pakistan

S. Ilyas · S. Naz
Department of Biotechnology, Lahore College for Women University, Lahore, Pakistan

S. Javad (ed.), *Nanoagronomy*, https://doi.org/10.1007/978-3-030-41275-3_2

can be synthesized by geogenic, biogenic and pyrogenic procedures (Nowack and Bucheli 2007). The use of nanoparticles in different devices and procedures depend upon their size as smaller the particles are, larger is the surface area per mass unit (Sozer and Kokini 2009; Ma et al. 2011). Other than this, their reactivity with other materials is also influenced due to their shape, surface covering, parent material or ion type (Khodakovskaya et al. 2012; Ranjan et al. 2014; Maddineni et al. 2015; Dasgupta et al. 2016; Jain et al. 2016). When we talk about the biogenic nanoparticles (nanoparticles formed by plants, microbes or fungi), their properties are also affected by the type of organism used. These uncommon properties may result in unique destiny and practices than their partners. It is the main reason for the extensive research going on types of nanoparticles, their use and their effects when released to the environment. This increased research on nanoparticles has also left an impact on botanists and agronomists to establish a research view point about the physiology of plants and plant pathogens in relation to the unique properties of these nanoparticles (Siddiqui et al. 2015).

2.2 Plant Tissue Culture

In fast developing field of biotechnology, plant tissue culture has taken lead as the most promising area of application of biotechnological tools for agriculture, today and tomorrow. Plant tissue culture is a combination of *in vitro* techniques which can produce thousands of plants in lab conditions in very lesser time period. It is done by providing the artificial nutrient medium, sterilization and artificial environment to the plants. It includes micropropagation, callus cultures, somatic hybridization, hybrid cultures, protoplast cultures and cell suspension cultures. The applications range from mass propagation of agricultural crops, horticulture plants, trees and other commercial plantations. It also finds applications in industry where plant products are being used like pharmaceutical industry. Plant tissue culture is also a viable tool for germplasm conservation like synthesis of artificial seeds and cryopreservation. Plants are a great source of different biologically active secondary metabolites that promisingly take part in subsistence of plants in a particular environment. *In vitro* tissue culture has found to be beneficial for the synthesis of secondary metabolites. In the era of ever demanding development of pharmaceutical industry, plant cell tissue culture holds great promise for controlled production of myriad of useful secondary metabolites.

The *in vitro* culture has an extraordinary job in practical and competitive agribusiness, forestry and pharmaceutical industry. It has also been effectively applied in plant rearing for quick production of improved plants. In plant rearing, tissue culture has become a fundamental part. At present, plant tissue culture has been incredibly advanced. Micropropagation is a quick, top notch, ailment free and uniform method of creating planting stock. Plant creation can be completed during the time regardless of climate and season.

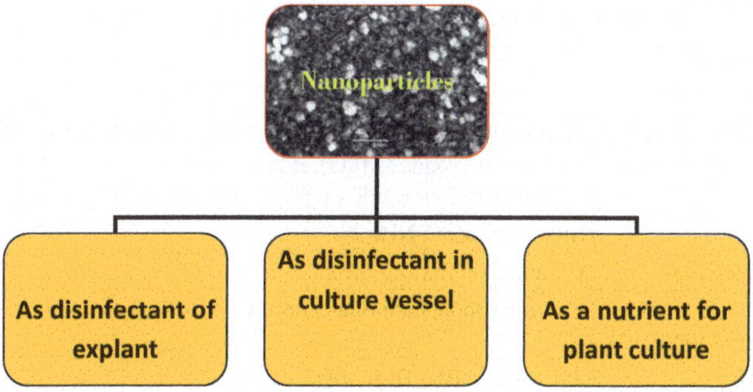

Fig. 2.1 Sterilization of explants using nanoparticles

Another potential job of tissue culture is the production of transgenic plants in an effective way. Plant tissue culture provides naked cells of callus, cell suspension cultures, embryos and protoplast to play with them at gene level more easily and precisely. With the help of these designed NPs DNA and proteins are transferred into plants to make transgenic plants (Park et al. 2008).

One of the utilization of tissue culture is in the preservation of germplasm. The germplasm of uncommon and endangered plants just as new and exceptional ones can be conserved *in vitro* through tissue culture. When some malady or climatic catastrophe clears out harvests, then plant tissue culture techniques have more capacity to replace the loss due to its rapid duplication rate. Its capacity to produce disease free plants also plays its role. The loss of hereditary assets is a typical story when germplasm is held in field quality banks. *In vitro* storage utilizing plant tissue culture apparatuses and cryopreservation are being proposed as answers for the issues innate in field quality banks. By these methods the future ages will have the option to approach hereditary assets for basic customary reproducing programs, or for the more unpredictable hereditary changes.

2.2.1 Nanoparticles in Tissue Culture

Tissue culture of plants is the center of plant sciences, which is significant for protection, mass spread, hereditary control, creation of bioactive mixes and plant improvement. There are numerous reports in plant tissue culture that showed contributions of nanotechnology in positive way. But there are two main roles of nanoparticles in plant tissue culture i.e., they may act as nutrients or they may be used as disinfectant or both (Fig. 2.1).

Iron and magnesium nanofertilizers usage essentially enhanced the quantity of seeds per pod and the protein content of the seeds in dark grown peas (Siddiqui and Al-Whaibi 2014). In tobacco cells, protoplast and leaves Au-topped mesoporous silica nanoparticles were found to be helpful in transfer of DNA (Delfani et al.

Table 2.1 Role of nanoparticles in plant tissue culture

Sr No	Tissue culture technique	Type of nanoparticles
1	Sterilization of explants	Aluminum oxide (Al_2O_3), CuO, iron oxide (Fe_3O_4), gold (Au), magnesium oxide (MgO), nickel (Ni), silver (Ag), silicon (Si), SiO_2, titanium dioxide (TiO_2) and ZnO NPs
2	Callogenesis	Silver and ZnO NPs
3	Organogenesis	Cu and Co NPs
4	Genetic transformation	Calcium phosphate (CaP) and Au NPs
5	Secondary metabolite production	Ag, Al_2O_3 NPs, TiO_2 NPs

2014). Similarly, nanoparticles of copper oxide (CuO) and zinc oxide (ZnO) increased the quantity of phenolics, anthocyanins, flavonoid, tannins, glycyrrhizin in the licorice seedlings (Torney et al. 2007). Silver and silica nanoparticles were accounted for to have antimicrobial action against disease causing agents in plants. These NPs controlled the mold in squash plants (Oloumi et al. 2015). Generally different nanoparticles (NPs) are being utilized to improve tissue culture response of explant. Different nanoparticles used in this regard are tabulated as Table 2.1.

2.2.2 *Nanomaterials and Surface Sterilization of Explants*

In tissue culture, microbial infection (both bacterial and fungal) is a significant issue. Bacterial infection appears to be a white, yellowish white or slimy growth on culture medium (Fig. 2.2) or on explant while fungal infection can be identified usually due to their colorful fruiting bodies. Microbial infection or sullying, even before induction, can stifle the whole procedure and productivity of the designed experiment. The research facility condition and even explants themselves are the good and rich source of contaminants. Surface cleansing of the explants is a significant advance preceding *in vitro* culture inception since microorganisms develop quicker in culture media than the explants, which are capable of culture commencement.

Generally, for the establishment of *in vitro* experiments, explants are gathered from their naturally growing areas or from nurseries. These explants are then surface sterilized. For this purpose different chemicals are used. A few disinfecting operators, for example, ethanol, bromine water (BW), hydrogen peroxide (H_2O_2), mercuric chloride ($HgCl_2$), sodium hypochlorite (NaOC1), calcium hypochlorite (CaOCl), silver nitrate ($AgNO_3$), fungicides and anti-infection agent are utilized to sterilize explants (Leifert et al. 1994). If the explant tissue is soft, then the use of higher concentration of disinfectant chemicals causes the death of explant tissue instead of or along with microbial spores. Nanoparticles can prove their utility here.

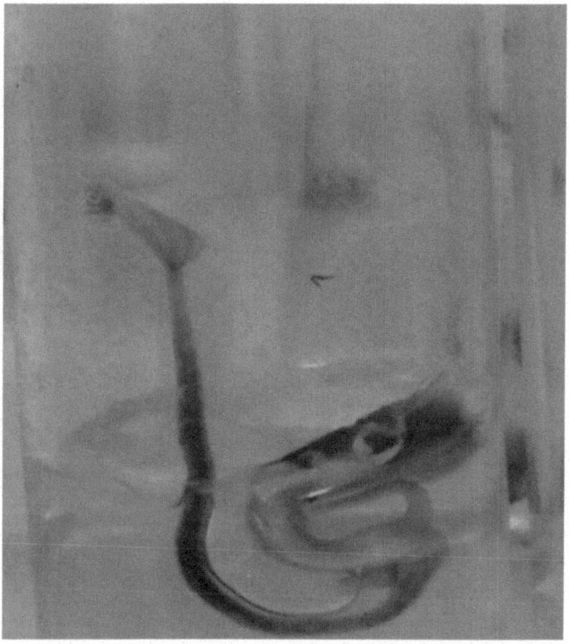

Fig. 2.2 Contamination of cultures

Utilization of nanoparticles (NPs) in plant tissue culture has effectively eliminated the microbial infection from explants. It exhibited positive effects on the establishment of organogenesis, callus, somatic embryos, and cell suspension cultures etc. Nanoparticles just minimized or finished the infection but did not affect the health of explant at its optimum conditions (Wang et al. 2016; Ruttkay et al. 2017). Reported literature has indicated that surface sterilization of explants with nanoparticles essentially decreases microbial infections from them.

Nanoparticles as antibacterial agent is not a new subject and it is one of the well established and settled topic. Different microbial agents were found to be eradicated as a result of utilizing metal and metal oxide nanoparticles (Tambarussi et al. 2015). Most famous nanoparticles being used are ZnO, Ag and TiO_2 (Rajput et al. 2018). Some other nanoparticles such as ZnO, Fe_2O_3, MgO, Al_2O_3, CuO, Ni, SiO_2, TiO_2 and many more are being used in plant tissue culture as sanitizing agents for explants and do not have any negative effect on explants. Though, seed germination, cotyledon and leaf endurance might be sometimes diminished if explants are treated with higher levels of Silver nanoparticles.

Removal of contaminants in plant tissue culture depends upon the type, size and dispersion of the nanoparticles. Few studies reported that the endurance and recovery of explants was increased with the application of NPs (Agista et al. 2018). This shows that there is need of research protocols to optimize the NPs concentration for various plant species to sterilize the targeted explant without any phyto-toxicity. The combined impact of sanitizing agents or antimicrobials and NPs can improve the viability of explants.

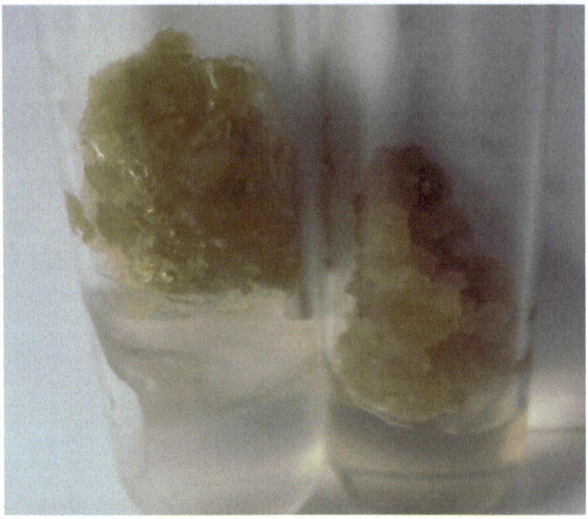

Fig. 2.3 Effect of ZnO NPs (left) on callus induction as compared to NPs free medium (right) from stem explant of *Dioscorea deltoidea.*

2.2.3 *Role of Nanoparticles in Callogenesis, Organ Induction, Shoot and Root Growth*

There are several reports that demonstrate callus formation, root and shoot generation as a result of nanoparticles for example, in *Tecomella undulata*, when silver nanoparticles were added to the medium in which stem explants were cultured the percentage of shoot induction, number of shoots and formation of callus was increased (Aghdaei et al. 2012). When *T. undulata* nodal explants were cultured on MS medium containing silver nanoparticles, number of shoots, percentage of shoot production and shoot length were considerably increased (Sarmast and Salehi 2016). Delayed senescence and enhanced survival rate by down regulation of TuACS gene was also observed by adding Ag NPs. Another reason of these positive effects of silver nanoparticles on organogenesis might be because of hindrance of ethylene creation. Zinc oxide nanoparticles (ZnO NPs) also have positive effects on callus proliferation (Fig. 2.3).

Adding nanoparticles to tissue culture medium influences proliferation of callus, multiplication of shoots, somatic embryogenesis and formation of roots by changing the activity of antioxidant enzymes, expression of genes and hindrance in ethylene generation. The authentic mechanism of the inhibitory or promotive impacts of nanoparticles on every parameter should be explored extensively. In *in vivo* experiments, effects of metal and metal oxide nanoparticles on plants have been studied. These NPs increase the morphogenetic capacity of explants acquired from different plants. However, effect of different doses of nanoparticles on various media like shoot induction and multiplication and media needs to be evaluated so that basic knowledge about nanoparticles application in tissue culture and their mechanism can be recognized.

2.2.4 Effect of Nanomaterials on Genetic Transformation

Genetic transformation is now the backbone of agronomy, used to produce new varieties with selected characters. Common methods of gene transfer to the plant cells, tissues and organs are agrobacterium mediated transformation, electroporation and particle bombardment. Many tissue culture products are utilized as source of plant tissue like callus cultures, somatic embryos or protoplasts. Generally electroporation is utilized to transfer genes into the protoplasts but it is not an easy job to isolate and regenerate the protoplast.

Nanoparticles also find a number of applications to ease the genetic transformation in plant tissue cultures. For example in protoplast isolation, to minimize the effects of cellular enzymes nanoparticles are used. DNA is transferred into the tobacco protoplast with the help of mesoporous silica nanoparticles by endocytosis. Similarly, biolistic gun is used to transfer gold coated silica nanoparticles, chemicals and DNA to the callus and leaves (Torney et al. 2007).

Carbon coated gold nanoparticles transmitted DNA into the *Oryza sativa, Leucaena leucocephala* and *Nicotiana tabacum* in contrast to standard gold particles utilizing a gene gun (Kumar et al. 2010). Carbon coated gold nanoparticles had lower measurement of gold and plasmid as compared to the business micrometer estimated gold particles. Plant cell injuries are almost insignificant; consequently recovery of plant tissue is easier and valuable.

To dispose off agrobacterium from cells, tissues and organs of different plant parts following co development nanoparticles require additional testing so as to give a proof. Poly (amidoamine) dendrimer nanoparticles is an innovative technique which efficiently transferred green-fluorescent protein (GFP)-encoding plasmid DNA into turf-grass cells (Pasupathy et al. 2008). By reforming pH of the medium and molar concentration of the dendrimer to the DNA transfection efficacy was enhanced. To harboring the GUS quality into *Brassica juncea*, Naqvi et al. (2012) used calcium phosphate (CaP) NPs for transferring pCambia 1301 vector. The best results were obtained from calcium phosphate NPs (80.7%) rather than *Agrobacterium tumefaciens* (54.4%) and DNA (8%). Similarly, in tobacco cells CaP NPs transferred pBI121 harboring GFP gene (Ardekani et al. 2014). In tobacco protoplasts and Arabidopsis roots mesoporous silica nanoparticles have been used to transmit plasmid DNA (Chang et al. 2013). In canola and carrot plants the conveying source of plasmid DNA to the protoplast and cells gold nanoparticles can be utilized. In plant nanobiotechnology, NPs mediated transmittance has gained extraordinary importance.

According to Ewais et al. (2015) the expansion of silver nanoparticles in culture media have been effective to activate morphology and anatomy of the callus by modifying the DNA and protein profile in the calli of *Solanum nigrum*. However, it is needed to check the broad scope applications of nanoparticles to upgrade somaclonal varieties.

2.2.5 In Vitro *Conservation*

Advances in plant biotechnology, particularly those related to *in vitro* culture and molecular biology, have likewise given useful tools to help and improve preservation of plant diversity. Biotechnological strategies have been utilized for conservation of rare, endangered, ornamental, medicinal and timberland species, permitting the preservation of pathogen free plant material, world class plants and hereditary assorted variety for short, medium and long term. *In vitro* protection is particularly significant for vegetatively proliferated and for non-conventional seed plant species. Besides, *in vitro* strategies offer a safe way to trade plant material internationally, collection of bulk material utilizing least space, permit supply of important material for wild populace recovery and encourage molecular investigations.

2.2.6 *Nanomaterials May Lead to Somaclonal Differences*

Somaclonal variation occurs when *in vitro* grown organs and plantlets are transformed. Basically, it is a result of changes in chromosomes (structure and number) and DNA (methylation, succession, initiation of transposable components and mitotic traverse) (Bairu et al. 2011; Sivanesan and Jeong 2012). Variations have a few valuable qualities like plant size, color of flowers, leaf variegation, natural product maturing, production of plant secondary metabolites and protection from biotic and abiotic stresses (Jeong and Sivanesan 2015). Phytotoxicity of nanoparticles at some levels has also been reported in few studies. Application of nanoparticles influences the mitotic activity, DNA stabilization, protein modification and DNA articulation in plants (Atha et al. 2012; Landa et al. 2012; Tripathi et al. 2017). MS medium containing carbon nanoparticles improved the ploidy level in *L. usitatissimum* calli. Amount of tetraploid cells was enhanced by supplementing carbon nanoparticles. DNA methylation was also higher in calli grown in carbon nanoparticles mediated medium. Effect of gold and silver nanoparticles on somaclonal variation in *L. usitatissimum* was reported by Kokina et al. (2017). Somaclonal variation was found to be increased in two calli and recovered shoot developed on media containing gold and silver nanoparticles. In *Solanum nigrum* calli silver nanoparticles quickens callus morphology and life structures by modifying protein and DNA profile (Ewais et al. 2015). To improve the somaclonal variety, additional experimentation is required to determine the larger scope of nanoparticles' application.

2.2.7 *Nanomaterials Cause Enhancement of Secondary Metabolites*

Plant tissue culture is also a good source of the commercial production of plant secondary metabolites and extensive research is going on. Depending on the plant and plant metabolite, different types of cultures are established. Some secondary metabolites are obtained from rooted cultures, some from callus cultures while some from cell suspension cultures. It is also a solid fact that by changing certain parameters of culture medium like culture components, growth regulators, light or dark conditions metabolite production can be improved (Jeong and Sivanesan 2015).

In plant tissue culture, addition of nanoparticles in medium may act as elicitor and supplement source. Utilization of Al_2O_3 nanoparticles (10–100 mg mL^{-1}) to tobacco cell suspension medium enhanced the phenolic content (Poborilova et al. 2013). Though, accumulation of phenolics in the cells was dose and time dependent. Nanoparticles in plant cell and organ cultures can be utilized as promising and efficient bioactive compound elicitors. For the production of secondary metabolite in plant tissue cultures additional knowledge is necessary to determine the capability of nanoparticles as elicitors.

2.3 Are Nanoparticles Toxic in Plant Tissue Culture?

Nanomaterials undoubtedly hold diverse properties, but it is quite challenging to work with unidentifiable compounds. Branch of nanotechnology dealing with harmful aspects of nanomaterials is known as Nanotoxicology. They have distinctive properties as compared to the larger counterparts; but they may possess additional toxic features, unlike the corresponding materials.

Although nanomaterials are sometimes made up of inactive components like gold, they become profoundly dynamic at nanometer measurements. Nanotoxicological studies are needed to evaluate the degree of risk associated with these nanoparticles. This risk evaluation may be based upon individual basis or global level. Nanotechnology has brought about a great deal of progressive changes, at the same time it is causing pollution. They are too small to be detected or studied easily. Although the term has been introduce as "The Nanopollution" but its aspects are not fully understood yet. But it could result in making another environmental catastrophe due to its undefined long lasting impacts. There is need to develop new policies to check size issues of nanoparticles. Up to this point, the poisonous effect of nanoparticles on plants has been generally assessed during *in vitro* seed germination and resultant development of seedlings. The impacts of nanopollution on human health are yet to be completely comprehended, making nanopollution one more man-made natural fiasco with dubious long term impacts really taking shape.

Scientists have extravagantly looked into the dangers of nanomaterials on plants (Miralles et al. 2012; Chichiricc'o and Poma 2015; Zaytseva and Neumann 2016;

Yang et al. 2017; Mishra et al. 2017). They propose that the nanomaterials added to the medium can prompt noteworthy and antagonistic impacts on viability of cells, organogenesis, shoot development, seed germination, explant survival and seedling growth. *In vitro* seed germination and seedling development of Alfalfa, grain, maize, rice, tomato and wheat was negatively affected by high dosages of carbon nanomaterials and metal NPs (Zaytseva and Neumann 2016).

Adding nanoparticles to cell suspension resulted in reduction of cell viability by varying expression of nucleic acid that induced DNA damage, disturbed synthesis of chlorophyll, and caused damage to cell membrane and leakage of electrolyte (Yang et al. 2017). Although, the uptake of nanoparticles by plants has not been evaluated, it has been reported that the uptake of nanoparticles by cells, tissues and organ cultures of plants is closely linked with uptake of moisture and nutrients from the medium (Lee et al. 2010). More elaborate dosage dependent studies are required for the identification of the preferably safe nanoparticles doses that have positive impacts on plant growth without any negative impacts on the plants as well as environment (Ruttkay et al. 2017).

Research is needed to assess the fate of nanoparticles during all the reaction steps of nanoparticles relating to their synthesis, characters and applications. Bioaccumulation, infiltration, and translocation of NPs in plants ought to be assessed in detail, explicitly for each type of NPs utilized, since each nanomaterial is one of a kind and the outcomes can't be summed up. This being a cross-disciplinary territory of research, suitable familiarity with the positive and negative parts of nanoparticles is compulsory for the client.

2.4 Future Projections

Without doubt, NPs have a ton to offer regarding different aspects of plant tissue culture. Starting from the principal stage, through sterilization, to separation of the callus, hereditary change, somaclonal variety and creation of secondary metabolites, NPs have exhibited their positive work. Nanotechnology offers boundless degree in different fields as it is a multidisciplinary subject. Although, a large number of nanoparticles are being utilized as antimicrobial agents but only few of them like Ag, TiO_2 and ZnONPs have been prevalently used to control microbial contamination in plant tissue culture. With materials like quantum dots, polymer dendrimers, graphene, carbon nanotubes, and nanowires, being created and all of them having antibacterial and antifungal properties, it is now time to discover new repositories of nanotechnology. The main issue in utilizing these nanomaterials unreservedly in living framework is the nanotoxicity perspectives and it should be surely known and thought upon. According to many studies nanoparticles can badly affect the explants and adversely influence the recovery capacity of explants. Whilst incorporating nanoparticles into culture medium clears the contamination and consequently beneficial for the germplasm, but their presence in the medium and wipes out microbial contaminants and consequently could be valuable for germplasm protection, their occurrence in the medium and their take-up by the plants needs to be observed. The impacts of NPs on induction of callus, organogenesis, shoot duplication, shoot extension, and establishing are additionally not all around reported. Also, the subculture impacts on callus expansion and shoot increase by addition of NPs should be considered in detail. Different nanoparticles have specific influence on secondary metabolite production in several plant species and the chemical, physical and biological properties of compounds obtained from plant cultures treated with nanoparticles also need to be investigated in detail.

2.5 Conclusion

Nanotechnology is rising as promising and well known field for the development of plants. To highlight the positive characteristics of nanoparticles and to minimize their harmful effects, more targeted research needed to be done to conserve the endangered plants using plant tissue culture.

References

Aghdaei M, Salehi H, Sarmas MK (2012) Effects of silver nanoparticles on *Tecomella undulata* (Roxb.) Seem. Micropropagation. Adv Hort Sci 26(1):21–24

Agista MN, Guo K, Yu Z (2018) A state-of-the-art review of nanoparticles application in petroleum with a focus on enhanced oil recovery. Appl Sci 8(6):871. https://doi.org/10.3390/app8060871

Ardekani MRS, Abdin MZ, Nazima N, Mohammed S (2014) Calcium phosphate nanoparticles a novel nano-viral gene delivery system for genetic transformation of tobacco. Int J Pharm Pharmaceut Sci 6(6):605–609

Aslani F, Bagheri S, Julkapli NM, Juraimi AS, Hashemi FSG, Baghdadi A (2014) Effects of engineered nanomaterials on plants growth: an overview. Sci World J 4: 641759., 28 pages https://doi.org/10.1155/2014/641759

Atha DH, Wang H, Petersen EJ, Clevela D, Holbrook RD, Jaruga P, Dizdaroglu M, Xing B, Nelson BC (2012) Copper oxide nanoparticle mediated DNA damage in terrestrial plant models. Environ Sci Technol 46(3):1819–1827

Bairu MW, Aremu AO, Van Staden J (2011) Somaclonal variation in plants: causes and detection methods. Plant Growth Regul 63:147–173

Bhushan R, Kumar R (2010) Enantioresolution of DL-penicillamine. Biomed Chrom 24(1):66–82

Chang FP, Kuang LY, Huang CA, Jane WN, Hung Y, Hsing YC, Mou CY (2013) A simple plant gene delivery system using mesoporous silica nanoparticles as carriers. J Mater Chem B 1:5279–5287

Chichiricc'o G, Poma A (2015) Penetration and toxicity of nanomaterials in higher plants. Nanomaterials 5:851–873

Dasgupta N, Shivendu R, Bhavapriya R, Venkatraman M, Chidambaram R, Avadhani GS, Ashutosh K (2016) Thermal co-reduction approach to vary size of silver nanoparticle: its microbial and cellular toxicology. Environ Sci Pollut Res 23(5):4149–4163

Delfani M, Firouzabadi MB, Farrokhi N, Makarian H (2014) Some physiological responses of black-eyed pea to iron and magnesium nanofertilizers. Commun Soil Sci Plant Anal 45:530–540

Ewais EA, Desouky SA, Elshazly EH (2015) Evaluation of callus responses of *Solanum nigrum* L. exposed to biologically synthesized silver nanoparticles. Nanosci Nanotechnol 5:45–56

Huang S, Wang L, Liu L, Hou Y, Li L (2015) Nanotechnology in agriculture, livestock, and aquaculture in China: a review. Agron Sustain Dev 35(2):369–400

Jain A, Shivendu R, Nandita D, Cidambaram R (2016) Nanomaterials in food and agriculture: an overview on their safety concerns and regulatory issues. Crit Rev Food Sci 58:297–317. https://doi.org/10.1080/10408398.2016.1160363

Jeong BR, Sivanesan I (2015) Direct adventitious shoot regeneration, *in vitro* flowering, fruiting, secondary metabolite content and antioxidant activity of *Scrophularia takesimensis* Nakai. Plant Cell Tissue Org Cult 23:607–618

Khodakovskaya MV, DeSilva K, Biris AS, Dervishi E, Villagarcia H (2012) Carbon nanotubes induce growth enhancement of tobacco cells. ACS Nanoparticles 6(3):2128–2135

Kokina I, Mickeviča I, Jermaļonoka M, Bankovska L, Gerbreders V, Ogurcovs A, Jahundoviča I (2017) Case Study of somaclonal variation in resistance genes *mlo* and *pme3* in flaxseed (*Linum usitatissimum* L.) induced by nanoparticles. Int J Genomics. https://doi.org/10.1155/2017/1676874

Kumar P, Fennell P, Robins A (2010) Comparison of the behavior of manufactured and other airborne nanoparticles and the consequences for prioritizing research and regulation activities. J Nanopart Res 12:1523–1530

Landa P, Vankova R, Andrlova J, Hodek J, Marsik P, Storchova H, White JC, Vanek T (2012) Nanoparticle-specific changes in *Arabidopsis thaliana* gene expression after exposure to ZnO, TiO_2, and fullerene soot. J Hazard Mater 241:55–62

Lee WM, Kim SW, Kwak JI, Nam SH, Shin YJ, An YJ (2010) Research trends of ecotoxicity of nanoparticles in soil environment. Toxicol Res 26(4):253–259

Leifert C, Morris CE, Waites WM (1994) Ecology of microbial saprophytes and pathogens in tissue culture and field-grown plants: reasons for contamination problems *in vitro*. Crit Rev Plant Sci 13:139–183

Ma X, Zhao Y, Liang XJ (2011) Theranostic nanoparticles engineered for clinic and pharmaceutics. Acc Chem Res 44:1114–1122

Maddineni SB, Badal KM, Shivendu R, Nandita D (2015) Diastase assisted green synthesis of size-controllable gold nanoparticles. RSC Adv 5:26727–26733

Miralles P, Church TL, Harris AT (2012) Toxicity, uptake, and translocation of engineered nanoparticles in vascular plants. Environ Sci Technol 46(17):9224–9239

Mishra S, Singh BR, Naqvi AH, Singh HB (2017) Potential of biosynthesized silver nanoparticles using *Stenotrophomonas* sp. BHU-S7 (MTCC 5978) for management of soil-borne and foliar phytopathogens. Sci Rep 7:45154

Naqvi S, Maitra AN, Abdin MZ, Akmal MD, Arora I, Samim A (2012) Calcium phosphate nanoparticles mediated genetic transformation in plants. J Mater Chem 22:3500–3507

Nowack B, Bucheli TD (2007) Occurrence, behavior and effects of nanoparticles in the environment. Environ Pollut 150(1):5–22

Oloumi H, Soltaninejad R, Baghizade A (2015) The comparative effects of nano and bulk size particles of CuO and ZnO on glycyrrhizin and phenolic compounds contents in *Glycyrrhiza glabra* L. seedlings. Ind J Plant Phys 20(2):157–161

Park EJ, Yi J, Chung KH, Ryu DY, Choi J, Park K (2008) Oxidative stress and apoptosis induced by titanium dioxide nanoparticles in cultured BEAS-2B cells. Toxicol Lett 180:222–229

Pasupathy K, Lin S, Hu Q, Luo H, Ke PC (2008) Direct plant gene delivery with a poly (amidoamine) dendrimer. Biotechnol J 3(8):1078–1082

Poborilova Z, Opatrilova R, Babula P (2013) Toxicity of aluminium oxide nanoparticles demonstrated using a BY-2 plant cell suspension culture model. Environ Exp Bot 91:1–11

Rai M, Ingle A (2012) Role of nanotechnology in agriculture with special reference to management of insect pests. Appl Microbiol Biotechnol 94(2):287–293

Rajput VD, Minkina T, Arvind B, Movsesyan HS (2018) Effect of zinc oxide nanoparticles on soil, plants, animals and soil organisms: a review. Environ Nanotechnol Monitor Manag 9:76–84

Ranjan S, Nandita D, Arkadyuti RC, Melvin SS, Chidambaram R, Rishi S, Ashutosh K (2014) Nanoscience and nanotechnologies in food industries: opportunities and research trends. J Nanopart Res 16(6):2464. https://doi.org/10.1007/s11051-014-2464-5

Ruttkay BN, Olga K, Lukas N, Vojtech A (2017) Nanoparticles based on essential metals and their phytotoxicity. J Nanobiotechnol 15(1):33. https://doi.org/10.1186/s12951-017-0268-3

Sarmast MK, Salehi H (2016) Silver nanoparticles: an influential element in plant nanobiotechnology. Mol Biotechnol 58(7):441–449

Siddiqui MH, Al-Whaibi MH (2014) Role of nano-SiO_2 in germination of tomato (*Lycopersicum esculentum*seeds Mill.). Saudi J Biol Sci 21(1):13–17

Siddiqui MH, Al-Whaibi MH, Mohammad F (eds) (2015) Nanoparticles and their impact on plants. Nanotechnology and plant sciences. Springer, New York

Sivanesan I, Jeong BR (2012) Identification of somaclonal variants in proliferating shoot cultures of *Senecio cruentus* cv. Tokyo Daruma. Plant Cell Tissue Organ Cult 111:247–253

Sozer N, Kokini JL (2009) Nanotechnology and its applications in the food sector. Trends Biotechnol 27(2):82–89

Tambarussi EV, Rogalski M, Nogueira FTS, Brondani GE, De Martin VF, Carrer H (2015) Influence of antibiotics on indirect organogenesis of teak. Ann For Res 58(1):177–183

Torney F, Trewyn BG, Lin VS, Wang K (2007) Mesoporous silica nanoparticles deliver DNA and chemicals into plants. Nat Nanotechnol 2:295–300

Tripathi DK, Singh S, Singh VP, Prasad SM, Dubey NK, Chauhan DK (2017) Silicon nanoparticles more effectively alleviated UV-B stress than silicon in wheat (*Triticum aestivum*) seedlings. Plant Physiol Biochem 110:70–81

Wang D, Mansisidor A, Prabhakar G, Hochwagen A (2016) Condensin and Hmo1 mediate a starvation-induced transcriptional position effect within the ribosomal DNA array. Cell Rep 14(5):1010–1017

Yang Y, Qin Z, Zeng W, Yang T, Cao Y, Mei C, Kuang Y (2017) Toxicity assessment of nanoparticles in various systems and organs. Nanotechnol Rev 6(3):279–289

Zaytseva O, Neumann G (2016) Carbon nanomaterials: production, impact on plant development, agricultural and environmental applications. Chem Biol Technol Agric 3:17. https://doi.org/10.1186/s40538-016-0070-8

Chapter 3
Nanotechnology and Abiotic Stresses

Sumera Iqbal, Zainab Waheed, and Alia Naseem

3.1 Introduction

The environmental factor restricting plant growth, vitality and fertility is known as abiotic stress. These include salinity, drought, heavy metal, extremely low or high temperatures, high UV radiation or low light, soil alkalinity or acidity and nutrient deficiency etc. Sensitivity to abiotic stress is exhibited by all major crops. The agricultural records reveal that tolerance of plants to stress has not been improved by crop cultivation through selection of parameters e.g. biomass accumulation, faster growth and seed and fruit yield. During stress conditions, vegetative growth and reproductive development of plants is restricted as a first response to stress, utilizing all plant energy resources and metabolic precursors to endure the effects of stress (Boscaiu et al. 2008).

An estimation of Food and Agriculture Organization (FAO 2017) depicts that 70% rise in agricultural productivity will be required in developing countries by 2050 since population of world will be increased to 9.1 billion by 2050. Certainly, the environmental factor such as climate change or global warming contributes to reduction in agricultural production. The effect of climatic change on agricultural productivity is crucial as it can influence the growth and development of plants through temperature changes and higher levels of precipitation and carbon dioxide. But a significant decrease in crop productivity is caused by factors other than global warming such as soil salinity, water shortage, floods, soil erosion and nutrient deficiency. Therefore improvement of crop varieties by genetic resources or genetic engineering and efficient measures for nutrient management and irrigation practices will be required for increased food production satisfying the growing population. Mostly in developing countries, the recent technologies related to agriculture focused at rising productivity generally do not consider the environmental aspect.

S. Iqbal (✉) · Z. Waheed · A. Naseem
Department of Botany, Lahore College for Women University, Lahore, Pakistan

S. Javad (ed.), *Nanoagronomy*, https://doi.org/10.1007/978-3-030-41275-3_3

Simultaneously, in each year, with the elevating cost of primary resources, the cost of irrigation and fertilizers rises which are required for administration in agricultural production. In this concern, to super-scribe the problem of elevation in adequate environmental management and productivity of agriculture under stress conditions, modern environmentally friendly approaches that do not demand big expenditures are required. In such circumstances, those approaches should be built on stimulating the adaptation capacity of plant (Kang et al. 2009).

Nanotechnology is going to be an emerging solution for agriculture under abiotic stresses (Hatami et al. 2016). In the previous decades nanobiotechnological applications have attained attraction of the researchers in the field of agriculture. Although toxic effects of nanoparticles on plants have been discussed many times however very little work has been done in explaining the mechanism of their action on plant growth and development. The transformation of materials to nano level not only brings about changes in their biological, chemical and physical properties but also influences catalytic characteristics (Manzer et al. 2015). The atomic aggregations having dimension (at least one) between 1 and 100 nm are called nanoparticles (Ball 2002). Nanoparticles are characterized by distinctive physical and chemical characteristics which are different from their bulk forms. Under abiotic stress (Fig. 3.1), yield of crops can be enhanced by the application of nanoparticles (Tamer et al. 2018). This chapter encompasses some recent research endeavors made in this respect.

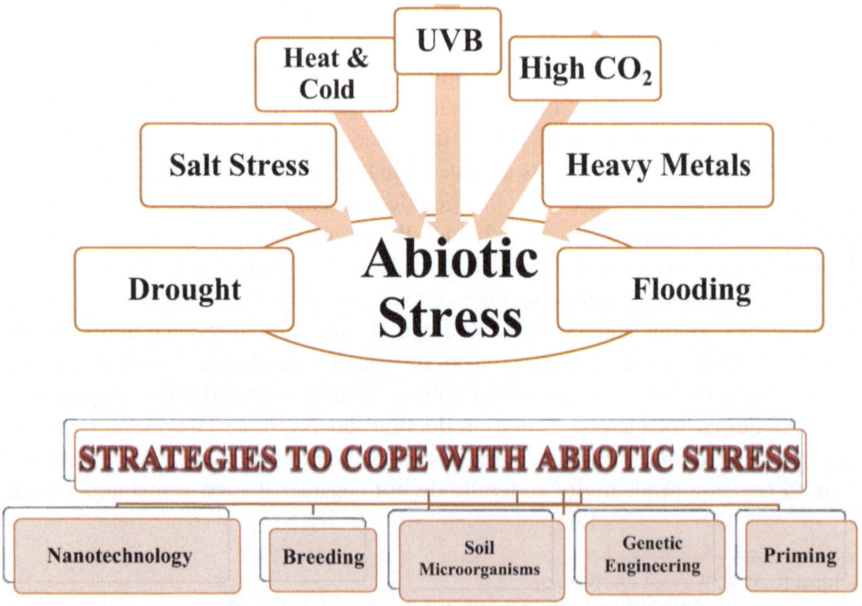

Fig. 3.1 Types of Abiotic stresses and possible strategies to cope with them

3.2 Drought

Drought is a major abiotic challenge that limits agricultural production. Plants face drought stress either when the supply of water to roots is lowered or when the rate of transpiration from leaves becomes very high. These two situations often coincide under semiarid and arid environments. Drought is a notable factor restraining production of crop in such dry areas. It is assumed that in the first period (quarter) of the twenty-first century, around 1.8 billion mankind will have to suffer entire shortage of water and about 65% of the people will live in circumstances of preferential shortfall of water (Nezhadahmadi et al. 2013).

Different nano particles have been utilized and found to have positive effects on plant growth and yield by modifying their physiological mechanisms. For example in a recent study, the colloidal suspension of Zn, Cu-nanoparticles were found to have a beneficial impact on morphometric indices. An anti-oxidative/pro-oxidative balance of leaves was noticed to be greater in steppe ecotype (Acveduc) seedlings and smaller in forest-steppe ecotype (Stolichna) seedlings under moisture deficit conditions. Under the influence of nanoparticles application under drought there was a decline in the quantity of Thiobarbituric acid reactive substances (TBARS) and increase of antioxidative enzyme activity (catalase and SOD) that designate the elevation of antioxidative activity of plant. And the impact of the colloidal Zn and Cu nanoparticle solution which caused a varying ratio of chlorophyll content in the leaves (Chl *a*/Chl *b*), in addition with increase of carotenoid contents in the leaves, was a demonstration of adaptation of plant to drought conditions. Moreover, variation in morphometric indices of plant, like relative water content of leaf and leaf area are the outcomes of the initiation mechanism of adaptation of plants to the conditions of drought (Astrid et al. 2018). In another research under different levels of moisture stress, the biochemical and physiological reactions of seedlings of hawthorn to a mixture of different amounts of nanoparticles (NPs) of silica were studied. Studies were carried out under moisture deficit condition and positive impacts of pre-treatment of nanoparticles on physiological parameters were noted. This research implicates that Silica NPs play a positive part in sustaining critical biochemical and physiological functions, in the seedlings of hawthorn during drought stress. It was reported that harmful impacts of drought stress on parameters of yield can be alleviated by foliar spray of iron NPs and percentage of oil of cultivars of safflower and the components of yield can be improved (Davar et al. 2014). Positive impacts of Titanium dioxide during drought stress on wheat have been delineated as well. This application enhanced contents of starch and gluten in seeds of wheat. Results proposed that supplementation of titanium dioxide NPs at the level of 0.02% exhibited increase in different agronomic traits that are ear weight, plant height, ear number, harvest index, 1000-seed weight, seed number, biomass, final yield involving content of starch and gluten during drought conditions (Jaberzadeh et al. 2013). Application of silver NPs lessens the negative effects of water stress on germination parameters of lentil (Hojjat 2016). It was also outlined that analcite [$AlSi_2O_6$]-H_2O NPs application promotes elevation in the resistance of wheat and corn plants to drought conditions. Parameters like growth criteria of seedlings, germination of seed and photosynthetic pigments content increased, but water balance characteristics did not deviate so much from the normal condition, during water stress. Application of Analcite enhanced precise protective antioxidants (carotenoids, flavonoids) accumulation and in corn, activation of catalase during drought conditions, while proline accumulation was not concerned with the demonstration of the analcite's protective action (Nataliya et al. 2014).

3.2.1 NPs and Plant Microbe Interaction

At the moment, a small number of studies specified the positive impacts of nanoparticles on the interactions between plants and microbes. It was outlined currently that CuO nanoparticles foster the growth of enlarged root hairs near the root tip, and moreover ZnO nanoparticles elevated lateral root establishment in the wheat (*Triticum aestivum* L.) seedlings. These above mentioned responses appeared in the roots that are invaded by *Pseudomonas chlororaphis* O6 (*Pc*O6); a valuable bacterium, that was initially isolated from the wheat roots grown on calcareous soils under dryland farming. In seedlings of wheat, the tolerance induced by bacterium (*Pc*O6) to drought stress was not reduced by the nanoparticles. Additionally, growth of the plants colonized by *Pc*O6 bacterium with nanoparticles resulted in systemic elevation in the genes expression related with adaptation to drought stress. Elevated expression of genes in the shoots associated with metal toxicity was compatible with elevated levels of Zinc and Copper in the plants that are colonized by the bacterium namely *Pc*O6 grown with the nanoparticles. This research work describes that plants grown with ZnO or CuO nanoparticles manifested cross-protection from distinct challenges like drought stress and metal toxicity. Same group of researchers used model system of wheat, grown in sand for the purpose of studying responses of seedling to ZnO or CuO nanoparticles (applied at levels of 500 and 300 mg metal/kg) respectively. These nanoparticles did not decrease layered biofilms formation on seedling roots of wheat by a probiotic namely *Pseudomonas chlororaphis* O6, inducing tolerance against drought. The results demonstrated that plant growth with 300 mg/kg CuO nanoparticles alone decreased shoot water content by the rate of 12%, in addition varied the mechanical features of the tissue in comparison with control plants. After the conditions of drought for 6 days, shoots of 13 days seedlings were visibly straighter when seedlings were grown with copper nanoparticles in comparison to the plants grown without the application of copper. Growth of plants colonized by bacterium namely PcO6 with CuO nanoparticles encouraged lignification of the sclerenchyma in shoots, furthermore elevated nitric oxide accumulations in the root of wheat. Nitric oxide, a metabolite concerned with cell signaling in tolerance against drought stress. These above mentioned studies demonstrated that formulations carrying chosen nanoparticles may interact positively with probiotics of plant in stimulating tolerance to drought and robust tissues of plant (Jacobson et al. 2018) (Table 3.1).

Table 3.1 Some recent studies in which positive effects of nanoparticles on plants were observed under different abiotic stresses

Stress type	Plant	Nanoparticles used	Reference
Chilling	*Triticum aestivum* L.	Biogenic silver nanoparticles	Bhati-Kushwaha et al. (2013)
Drought	*Carthamus tinctorious* L.	Iron	Davar et al. (2014)
Drought	*Zea mays* L. and *Triticum aestivum* L.	Nanoparticles of analcite	Nataliya et al. (2014)
Waterlogging	(*Glycine max* L.)	Nano Al_2O_3	Mustafa et al. (2015)
Waterlogging	(*Glycine max* L.)	Nano Ag	Mustafa et al. (2015)
Drought	*Lens culinaris* L.	AgNPs	Hojjat (2016)
UV-B	*Triticum aestivum* L.	Silicon Nano particles	Kumar and Swati (2016)
Drought	*Sesasum indicum* L.	Iron oxide	Mostafa et al. (2016)
High CO_2	*Oryza sativa* L.	$nTiO_2$	Du et al. (2017)
Salinity	*Zea mays* L.	ZnO	Fathi et al. (2017)
Salinity	*Trigonella foenum-graecum*	Silver nano particles	Hojjat and Kamyab (2017)
High temperature	*Moringa oleifera*	AgNPs	Iqbal et al. (2017)
Salinity	*Vicia faba* L.	$nTiO_2$	Mojtaba and Lam-Son (2018)
High temperature	*Sorghum bicolor* (L.) Moench	Se-NPs	Djanaguiraman et al. (2018)
Heavy metals	*Triticum aestivum* L.	Zinc oxide	Hussain et al. (2018)

3.3 Waterlogging

Waterlogging happens when the soil is too much wet that there is inadequate oxygen in the space of pore, it is difficult for plant roots to respire sufficiently. Plants vary in their requirement for oxygen. There is no general measure for oxygen of soil that can distinguish conditions of waterlogging for all of the plants. Moreover, in the root zone, the demand of oxygen for plant will differ with its growth stages. Waterlogging stress generates deficiency of oxygen that leads to hypoxia and ethylene production in plants that can hinder growth of roots, permeability of roots and it also decreases subsistence of most plants besides wetland species like rice.

3.3.1 *Effects of Waterlogging*

Waterlogging declines the sites that can be cropped in wet years. When the sites are waterlogged shortly after seedling, the results are with low or no emergence and germination of plants; and the crops may have to be re-sown. The waterlogged sites are very vulnerable to water erosion and soil structure degradation (Kahlown and

Majeed 2003). Furthermore, when these waterlogged sites are cultivated, these are more weakened. Waterlogging happens over a vast area of the world, negatively influencing about 10% of the worldwide area of land (Setter and Waters 2003) and decreasing yields of crops by as much as about 80% (Shabala 2011).

Under waterlogging circumstances, soil gas exchange is critically hindered. Due to root and microbial respiration, this results in a notable reduction of free oxygen and carbon oxide accumulation (Bailey-Serres and Voesenek 2008). In roots, hypoxia stress arises when there is the depletion of free oxygen in the soil zone surrounding the roots. This leads to a shift from aerobic to anaerobic metabolism, in connection with considerable restrictions to synthesis of ATP (Barrett-Lennard 2003; Teakle et al. 2006). Furthermore, waterlogging also provokes a sudden decrease in the redox potential of soil, and results in very notable changes to the chemical profile of soil. Effects comprise of a changed accessibility of mineral substances, decline of sulfate ($SO^{2-}{}_4$), iron (Fe^{3+}) and manganese (Mn^{4+}) and elevated solubility of metals and moreover microbial anaerobic metabolism and induction of toxic compounds by roots of plant (Kozlowski 1997; Shabala 2011). Due to these variations, plants manifest changed membrane transport, lessened water potentials of leaf and stomatal conductance, increased senescence of root, declined shoot and root growth, and finally it leads to death of plant (Barrett-Lennard 2003) (Fig. 3.2).

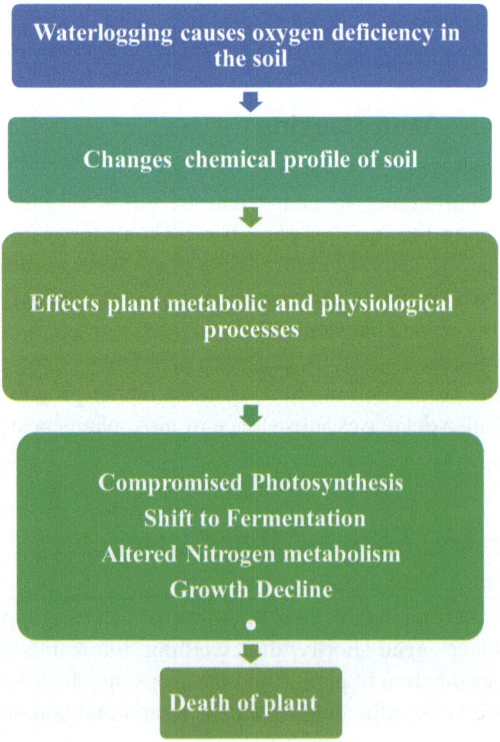

Fig. 3.2 How waterlogging affects plants?

3.3.2 Waterlogging and Nanoparticles

Flooding negatively influences the growth of different plant particularly the crops like wheat maize, vegetables and oil seed crops like soybean. And along with some others, silver nanoparticles are reported to elevate the growth during waterlogging stress. To understand this effect, a study was done to evaluate the response of soybean to application of silver nanoparticles under waterlogging stress and a proteomic (gel-free) technology was utilized for this purpose. Under flooding stress, the treatment of 2 ppm silver nanoparticles (size = 15 nm) facilitated the growth of soybean and increased growth of seedling of soybean. A subsequent study was conducted to examine the variations in the proteome profile after the treatment with silver nanoparticles during flooding. The abundance of proteins related to fermentation and glyoxalase II 3 were notably increased due to flooding stress; although, declined by treatment of silver nanoparticles. During treatment of silver nanoparticles, relatively decrease in the level of glyoxalase II 3 transcript entails that low by-products (cytotoxic) of glycolysis were created in soybeans that were exposed to silver nanoparticles in contrast to soybean that is flooded but not treated with NPs. Additionally, during flooding stress, up-regulation of genes of enzymes pyruvate decarboxylase and alcohol dehydrogenase and their down-regulation due to the treatment of silver nanoparticles was noticed. These might be linked to a change in metabolism towards normal cellular processing (Mustafa et al. 2015).

Saffron is a specie of the family namely Iridaceae. It is commercially utilized for aroma, food, coloring, therapeutic and cosmetic purposes. It is known that saffron is a good preventive and remedial agent for distinct cancers. As waterlogging is not favorable for the growth of other plants saffron is also affected negatively by this stress. This stress impacts many biochemical, morphological, and physiological processes, involving plant photosynthetic capacity, production rates of plant biomass, the amount of shoot and root growth, water relations, metabolism of carbohydrate, the structure of cell, nutrition and moreover influences gene expression (Crawford and Andle 1996) (Fig. 3.3).

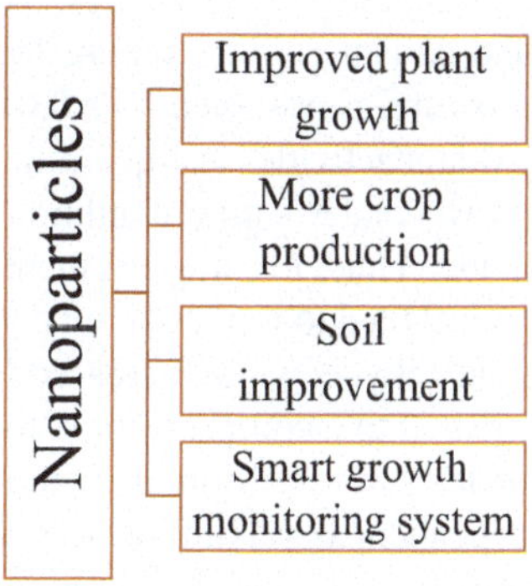

Fig. 3.3 Effects of nanoparticles

3.4 Salt Stress

Salt stress or soil salinity means excess of salts in soil and it is known as one of the major environmental evils, reducing crop productivity globally, thus has become a threat for sustainable agriculture. Salt affected soils have many problems like high Na content, poor porosity, waterlogging and loss of nutrients as well as the hydraulic constraints. Several approaches to manage salt-affected lands have been used which include the use of tolerant crops applying chemicals and nanomaterials. Nanotechnology is being utilized in different ways to overcome this problem of salinity. Different nanoparticles can be applied to plants for improving their growth under saline conditions. Nano-materials or nano-reclaimants can be effective in reclaiming salt-affected soils. Nano gypsum, nano calcium and magnesium compounds, were found to be some of efficient and readily manufactural reclaimants and are known to enhance soil hydraulic characteristics and stability (Mukhopadhyay and Kaur 2016; Patra et al. 2016).

Silver removes undesirable microorganisms in hydroponics systems and farmer soils. Moreover, it is being utilized as foliar spray against moulds rot, fungi and many other plant diseases. Additionally, silver is a considerable stimulator of plant growth, comprising of silicate, silver salt, and water soluble polymer to radioactive rays (Sharon et al. 2010). For ameliorating negative effects of salinity by silver nano particles on germination and growth of Fenugreek, an experiment was conducted by Hojjat and Kamyab (2017). Overall AgNPs application was found to be beneficial in improving salinity tolerance of Fenugreek seedlings and it was suggested that different defence mechanisms of plants against salt stress may be stimulated by applying nanoparticles. Silver ions like silver nanoparticles have been known to hinder ethylene action and in this way increase root growth (Monica and Cremonini 2009).

In a study, impacts of titanium dioxide nanoparticles ($nTiO_2$) on broad bean for alleviating soil salinity were evaluated. The influence of three $nTiO_2$ levels (0.01%, 0.02% and 0.03%) were evaluated on plant growth under stress. Significantly improved shoot length, leaf area and root dry weight were recorded due to application of $nTiO_2$ (0.01%) under normal conditions. The promoting effects were reported to be in the form of enhanced quantities of chlorophyll b, sugars, proline and antioxidant enzymes activities. Though under salt stress, proline level and enzymatic antioxidant activities were found to be higher and significant reduction in plant growth was observed. The 0.01% $nTiO_2$ application significantly enhanced the antioxidant activities and quantities of soluble sugars, amino acids and proline in plants grown under saline condition as compared to the plants subjected to salt stress alone. Thus, the improvement in the activities of antioxidant enzyme might have contributed to the observed decline in the contents of hydrogen peroxide and malondialdehyde, while increased levels of proline and some other metabolites has contributed to osmo-protection, together causing a significantly improved plant growth under salinity. Moreover, these positive effects were found to be concentration dependent as maximum effect was observed when with lowest concentration (0.01%

nTiO) was applied whereas intermediate response was noticed when 0.02% was applied and highest (0.03%) concentration was found almost ineffective under both control and salinity treatments (Mojtaba and Lam-Son 2018).

3.5 Temperature Stress

Temperature above or below the optimum for plants can be a stress factor. In plants, three categories of temperature stress can be identified based on the categories of stress like high, freezing or chilling. Temperature stress adversely affects plants by reducing growth, photosynthetic activities and germination rates ultimately leading to death of plants (Kai and Iba 2014). Studies are available which show positive influences of nanoparticles on plants under temperature stress.

For example an experiment was performed to investigate the influence of silver nano particles (AgNPs) on wheat growth under high temperature stress. The wheat plants were treated at their trifoliate phase with various levels of silver nano particles (25, 50, 75 and 100 mg/L). Heat stress with temperature of 35–40 °C was applied for 3 h per day for 3 days. According to the results, a decrease in number and length of roots, length of shoots, mass (fresh and dry) of plants, leaf number, leaf fresh weigh and dry mass and leaf area was observed when heat stress was applied alone. It was found that when AgNPs were applied to the plants they improved the plant growth parameters. At a concentration of 50 and 75 mg/L AgNPs improved the plant weight (fresh and dry), number and length of roots, length of shoots, number of leaves, fresh and dry mass of leaves and leaf area relative to control it can be concluded here that application of silver nanoparticles can protect wheat from high temperature stress. So from the above results, the conclusion can be made that AgNPs have ability to enhance growth of wheat under high temperature stress and can be used to protect this crop from negative effects of stress (Iqbal et al. 2017).

Selenium nanoparticles (Se-NPs) are also being investigated to see their potential to mitigate the negative effects of high temperature (HT) on plants. But there is inadequate information concerning the use of selenium nanoparticles (Se-NPs) for amelioration of high-temperature stress in agricultural crops. In a recent research, toxic, biological and physiological impacts of Se-NPs were studied in *Sorghum bicolor* (L.) Moench under high temperature. Se-NPs (10–40 nm in dimensions) were prepared and their characterization revealed their nanocrystalline nature. The translocation study carried out in sorghum plants revealed the root to shoot movement of Se-NPs. In sorghum, activation of antioxidant defense machinery with elevated activities of antioxidant enzymes was observed under high temperature stress Se-NPs (10 mg/L) was applied as foliar spray at the booting stage of plants. Moreover, concentration of oxidants was reduced by Se-NPs application. Se-NPs contributed to elevated content of unsaturated phospholipids. Higher germination percentage of pollen was observed by Se-NPs under HT stress resulting in pronounced increase in seed production. Compared to optimum temperature treatment

(32/22 °C), Se-NPs enhanced antioxidant enzyme activity and reduced content of oxidants efficiently under high temperature (HT) stress (38/28 °C).

It can be inferred that application of Se-NPs to sorghum plants can protect them against HT stress by stimulating antioxidative defense system. The investigations proved the significant effects of Se-NPs on physiological behavior of sorghum plants under HT stress. It is evident from the study that Se-NPs have potential to protect sorghum plants against HT stress. Moreover, this research provides first evidence of stimulation of antioxidant defense system by foliar spray of Se-NPs to sorghum plants grown under heat stress. The observed phospholipid content indicated that Se-NPs stabilized the thylakoid membrane composition and integrity during HT. The non-toxic nature of Se-NPs was confirmed by the response of pollen functions. It is suggested by researchers that investigations on Se-NPs in plants must be extended to understand their mode of action and metabolism. Extensive research work will be required to study the physiological and biochemical aspects of Se-NPs thoroughly and role of genes responsible for their uptake, assimilation and transport (Djanaguiraman et al. 2018).

Chilling temperatures are known to affect plant growth and crop productivity, leading to significant crop losses. In a research, biogenic nanoparticles (BioNPs) which were synthesized by using *Tridax procumbens* were utilized as a priming agent to stimulate the anti-oxidative mechanism, and to observe the effects of these nano particles on seed germination and seedling growth attributes of wheat. It was found that BioNPs significantly increased the activities of super oxide dismutase and other antioxidants (Bhati-Kushwaha et al. 2013). This study indicates the potential of nanopriming in boosting the antioxidant mechanism of plants which is very important under abiotic stresses.

3.6 Heavy Metal Stress

Metal contamination is becoming common in the world, among heavy metals, lead, chromium, copper mercury and cadmium are major environmental pollutants, especially in areas with greater anthropogenic pressure. A few of these metals, like Cu, Mn, Co, Zn and Cr are essential plant nutrients which are required by plants in trace amounts. But when metals are bioavailable and are present in excess amount, they become toxic to plants (Nagajyoti et al. 2010).

Accumulation of heavy metals in soils is a problem of great concern, in agricultural field particularly, due to the adverse effects on food security, crop production and marketability. They cause toxicity and destroy the environmental fitness of soil biota. Plants growing in the soils polluted with heavy metals show altered metabolism, reduction in growth, lesser biomass production and accumulation of metal. Physiology and biochemistry of plants is adversely affected by these metals. Research about toxicity and ways which can enhance tolerance in metal-stressed plants is required due to the growing metal pollution in the world.

3.6.1 Effects of Nanoparticles Under Heavy Metal Stress

There are reports in which nanoparticles were found to help in reducing the harmful effects or absorption/uptake of heavy metals. A research was done to investigate the impact of waste water (containing heavy metals) and nano-TiO_2 (2–6 nm) on growth of seedlings of maize. The nano-TiO_2 suspension either in autoclaved waste water or deionized water were applied at the levels of (100, 50 and 25) mg/L during *in vitro* experiments. Waste water analyses demonstrated that it was not acceptable for purposes of irrigation because it had elevated heavy metals content (Zn, Cu, Fe, Mn, Cr and Cd) that were above acceptable amounts for irrigation. The elevated amount of heavy metals in waste water and nano-TiO_2 at the level of (100 mg/L) markedly hindered germination of seed, growth of seedling and moreover caused phenolics accumulation in maize. Nano-TiO_2 application at the level of 25 mg/L significantly elevated dry and fresh weight of shoot, fresh and dry weight of root, area of root, carotenoids and chlorophyll (a and b) contents ($p < 0.05$). Adverse impacts of waste water on growth parameters of maize were markedly alleviated by nano-TiO_2 at the level of 25 mg/L. The application of waste water with nano-TiO_2 at the level of (25 mg/L) is suggested before its usage for the purpose of agriculture. The treatment of heavy metal contaminated water with nano-TiO_2 at the level of 25 mg/L showed advantageous impacts on maize growth features. But, the elevated amount of nano-TiO_2 was toxic to plants. This suggested that the application of nano-TiO_2 at the level of (25 mg/L) has no negative impacts on the growth of maize and can be suggested for future implementation to maize during irrigation with waste water which was polluted with heavy metals toxicity (Yaqoob et al. 2018).

A study was done to investigate the effects of magnetic (Fe_3O_4) nanoparticles (nano-Fe_3O_4) in alleviating of the heavy metal (Pb, Zn, Cd and Cu) toxicity in wheat seedlings. Application of magnetic (Fe_3O_4) nanoparticles (2000 mg/L) in the 1 mM solution of each heavy metal significantly declined the growth inhibition and stimulated protective mechanisms to decrease oxidative stress which was induced by heavy metals in the seedlings of wheat. The positive effects of nano-Fe_3O_4 under stress of heavy metals could be caused by the increase in the activities of antioxidant enzymes. Their mitigating effect was confirmed by the reduction in MDA content. The alleviating effects of nano-Fe_3O_4 were considered to be associated with their adsorption potential of heavy metals which could be due to a different electrostatic attraction between heavy metal ions and adsorption sites which are negatively charged (Konate et al. 2017).

Effectiveness of iron oxide nanoparticles (Fe_3O_4 NP) in reducing harmful effects of arsenic in *Brassica juncea* was observed in an experiment. Decrease in the plant stress-related parameters was noted which might be due to restricted absorption of Arsenic by the plant in the presence of Fe_3O_4 NP (Praveen et al. 2018). Zinc oxide nano particles were reported to increase the wheat chlorophyll content, gas exchange attributes, antioxidant enzymes, zinc uptake and yield but reduce the Cd concentration under cadmium stress (Hussain et al. 2018).

3.7 UV Stress

Over northern and southern hemispheres, depletion of ozone will presumably elevate (UV)-B radiation (285–315 nm), that will influence crop plants in various ways. The most common symptoms are found in UV-sensitive plants like reduction of growth; reduced photosynthesis and moreover decreased biomass are observed in plants that are UV-sensitive. This has been studied when UV-B stress is applied with artificial white light, or sometimes with solar light. In photosynthesis, under intensely high UV-B circumstances, the molecular targets for action of UV are probably Photosystem II reaction center area. But on the other hand, Photosystem I is less sensitive. In rice and soybean cultivars, reductions of yield have sometimes been noticed in plants that are sensitive to increased UV-B, but on the other hand, yield enhancement have also been noticed in some crops. Current researches while working on various plant cultivars and species; are impairing solar UV-B by special filters of UV and demonstrating reductions in yield and relative growth in cultivars that are sensitive to UV. This is caused by a decrease of development of plant and flowering during increase solar UV-B radiation. Investigation with trees reveals a decrease susceptibility to UV-B because of their strong structure of leaf and the assemblage of UV absorbing compounds, which are common response in various species of plants. A multitude of genetic, morphological and physiological outcomes are assumed for single species of plant and for ecosystems with elevated UV-B radiation, as assumed by ozone depletion scenarios (Tevini 2004).

Nanoparticles are gaining the attention of plant scientists in the recent years. Particularly among nanoparticles the role of silicon in minimizing abiotic as well as biotic stresses is familiar. UV-B radiation was reported to cause extreme impact on growth of seedlings of wheat, which was associated with reduced photosynthetic performance and changed vital structures of leaf. UV-B enhanced the amounts of superoxide and hydrogen peroxide and moreover elevated leakage of electrolyte and peroxidation of lipid peroxidation. Activities of antioxidant enzymes like ascorbate peroxidase (APX) and superoxide dismutase (SOD) were declined by UV-B but guaiacol peroxidase and catalase and, moreover non-enzymatic antioxidants were increased by UV-B. The research reveals that silicon nanoparticles might protect seedlings of wheat via activating antioxidant defense system mediated with nitric oxide, which successively counterbalance ROS-provoked photosynthesis damage. Moreover, silicon nanoparticles revealed to be more affective in declining UV-B stress in comparison to silicon, which is linked to its increase attainability to seedlings of wheat (Kumar and Swati 2016).

3.8 Carbon Dioxide (CO_2) Stress

Increased amount of carbon dioxide is having impact on climate at global level. Increasing carbon dioxide levels have great direct impacts on the physiology, chemistry and growth of plants, irrespective of any direct impacts on climate (Ziska 2008). As photosynthesizing organisms, plants take up CO_2, chemically declining the carbon from the atmosphere. This results in gaining of stored chemical energy for the plant but also supplies the carbon skeleton required for building molecules which makeup up plant structure. Photosynthesis is the main metabolic process of plants, and rise in the levels of carbon dioxide for photosynthesis can have great influence on growth and physiology of plant but up to an optimum concentration. High concentration of carbon dioxide causes a decline in photosynthesis in many plants. There is also indication from the past research that a sudden increase in carbon dioxide caused an increased damage to several varieties of plants species. Higher concentrations of carbon dioxide also decrease the nutritional quality of some staples, like wheat. Concerns have also been developed regarding the risks associated with nanomaterials, as it is very possible that nanomaterials/nanoparticles can be released to the environment and result in damaging effects to organisms. Carbon dioxide (CO_2) is one of the main greenhouse gases. The level of CO_2 is increasing day by day and is becoming a serious of environmental issue, especially for agricultural crops. A study was performed which showed that when seedlings were treated with NPs under high CO_2 stress, most of NPs were accumulated on the surface of roots (Miralles et al. 2012) which caused a decline in root hydraulic conductivity and thus reduced water availability and transpiration rate of the plants and repressed plant growth. Compared to the optimum CO_2 levels, wheat seedlings which were treated with the higher level of CO_2 showed higher root mass and lateral roots development. In a recent study effects of elevated CO_2 levels on toxicity of $nTiO_2$ (0, 50, and 200 mg/kg) against plants and microbes in a paddy soil system were evaluated. Results of this study showed that $nTiO_2$ did not cause toxicity in rice plants grown at the normal CO_2 level (370 μmol/mol), but the higher CO_2 concentration (570 μmol/mol) of $nTiO_2$ resulted in significant reduction of plant biomass, and grain yield. Moreover, at the high CO_2 levels, $nTiO_2$ caused increase in the accumulation of Ca, Mg, Mn, P, Zn, and Ti, but lowered fat and total sugar content of grains. These treatments also altered the community composition of soil microbes. Overall, this study indicates that rise in CO_2 levels would alter the effects of $nTiO_2$ on the nutritional composition of crops and function of soil microbial populations, with unidentified consequences for future economics and human health (Du et al. 2017).

3.9 Conclusion

No doubt, nanotechnology is an emerging field and has potential applications in diverse areas like electronics, energy, medical, and biological science. However, the use of nanotechnology and nanoparticles in the field of agriculture is still at initial stage particularly under stressful environments. Though in the recent years many preliminary studies related to this subject are performed which have given hope to the agricultural scientists that nanotechology can be used for the improvement of crop production in normal as well as stressful conditions. But, more studies are required to understand the mode of action of NPs, and their influence on the gene expression in plants. Furthermore increased utilization of nanotechnology in agriculture can result in the release of nanoparticles to the environment, so this aspect should not be ignored. Their effects should to be assessed in depth to avoid potential risks to the environment. At present only few studies are available on the phytotoxic effects of nanoparticles. So it is very important to optimize the size and levels of nanoparticles along with understanding the interaction between plants and nanoparticles before their applications in the fields so that their possible negative impacts can be overcome and they can be effectively used as a cost effective way to mitigate harmful effects of abiotic stresses on plants.

References

Astrid J, Stephanie D, Matthew P (2018) Interactions between a plant probiotic and nanoparticles on plant responses related to drought tolerance. Ind Biotechnol 14:3

Bailey-Serres J, Voesenek LACJ (2008) Flooding stress: acclimations and genetic diversity. Annu Rev Plant Biol 59:313–339

Ball P (2002) Natural strategies for the molecular engineer. Nanotechnology 13:15–28

Barrett-Lennard EG (2003) The interaction between waterlogging and salinity in higher plants: causes, consequences and implications. Plant Soil 253:35–54

Bhati-Kushwaha H, Kaur A, Malik CP (2013) The synthesis and role of biogenic nanoparticles in overcoming chilling stress. Indian J Plant Sci 2(4):54–62

Boscaiu M, Lull C, Lidon A, Bautista I, Donat P, Mayoral O, Vicente O (2008) Plant responses to abiotic stress in their natural habitats. Bull UASVM Hortic 65(1):53–58

Crawford RMM, Andle RB (1996) Oxygen deprivation stress in a changing environment. J Exp Bot 47:145–159

Davar FZ, Arash R, Amir H (2014) Evaluation the effect of water stress and foliar application of Fe nanoparticles on yield, yield components and oil percentage of safflower (*Carthamus tinctorious* L.). Int J Adv Biol Biomed Res 2(14):150–159

Djanaguiraman M, Belliraj N, Bossmann SH, Prasad V (2018) High-temperature stress alleviation by selenium nanoparticle treatment in grain sorghum. ACS Omega 3(3):2479–2491

Du W, Gardea-Torresdey JL, Xie Y, Yin Y, Zhu J, Zhang X, Ji R, Gu K, Peralta-Videa JR, Guo H (2017) Elevated CO_2 levels modify TiO_2 nanoparticle effects on rice and soil microbial communities. Sci Total Environ 578:408–416

FAO (2017) The future of food and agriculture – trends and challenges. FAO, Rome

Fathi A, Morteza Z, Shahram T (2017) Effect of interaction between salinity and nanoparticles (Fe_2O_3 and ZnO) on physiological parameters of *Zea mays* L. J Plant Nutr 40(19):2745–2755

Hatami M, Kariman K, Ghorbanpour M (2016) Engineered nanomaterial-mediated changes in the metabolism of terrestrial plants. Sci Total Environ 571:275–291

Hojjat SS (2016) The effect of silver nanoparticle on lentil seed germination under drought stress. Int J Farm Allied Sci 5(3):208–212

Hojjat SS, Kamyab M (2017) The effect of silver nanoparticle on fenugreek seed germination under salinity levels. Russ Agric Sci 43(1):61–65

Hussain A, Ali S, Rizwan M, Zia Ur Rehman M, Javed MR, Imran MS, Ali S, Chatha S, Nazir R (2018) Zinc oxide nanoparticles alter the wheat physiological response and reduce the cadmium uptake by plants. Environ Pollut 242B:1518–1152

Iqbal M, Raja NI, Mashwani ZUR, Hussain M, Ejaz M, Yasmeen F (2017) Effect of silver nanoparticles on growth of wheat under heat stress. Iran J Sci Technol A 43:387–395

Jaberzadeh A, Payam M, Hamid R (2013) Influence of bulk and nanoparticles titanium foliar application on some agronomic traits, seed gluten and starch contents of wheat subjected to water deficit stress. Not Bot Horti Agrobo 41(1):201–207

Jacobson A, Doxey S, Potter M, Adams J, Britt D, McManus P, McLean J, Anderson A (2018) Interactions between a plant probiotic and nanoparticles on plant responses related to drought tolerance. Ind Biotechnol 14:148–156

Kahlown MA, Majeed A (2003) Water resources situation in Pakistan: challenges and future strategies. PCRWR 3:17

Kai H, Iba K (2014) Temperature stress in plants [Abstract]. Encyclopedia of Life Sciences. https://doi.org/10.1002/9780470015902.a0001320.pub2

Kang Y, Khan S, Ma X (2009) Climate change impacts on crop yield, crop water productivity and food security. Prog Nat Sci 19:1665–1674

Konate A, He X, Zhang Z, Ma Y, Zhang P, Alugongo GM, Rui Y (2017) Magnetic (Fe_3O_4) nanoparticles reduce heavy metals uptake and mitigate their toxicity in wheat seedling. Sustainability 9:790

Kozlowski TT (1997) Responses of woody plants to flooding and salinity. Tree Physiol Monograph 1:1–29

Kumar TDS, Swati S (2016) Silicon nanoparticles more effectively alleviated UV-B stress than silicon in wheat (*Triticum aestivum*) seedlings. Plant Physiol Biochem 110:70–81

Manzer HS, Mohamed HAW, Mohammad F, Mutahhar YAK (2015) Role of nanoparticles in plants. Nanotechnol Plant Sci 19–35

Miralles P, Church TL, Harris AT (2012) Toxicity, uptake, and translocation of engineered nanomaterials in vascular plants. Environ Sci Technol 46:9224–9239

Mojtaba K, Lam-Son PT (2018) Titanium dioxide nanoparticles improve growth and enhance tolerance of broad bean plants under saline soil conditions. Land Degrad Dev 29(4):1065–1073

Monica RC, Cremonini R (2009) Nanoparticles and higher plants. Caryologia 62:161–165

Mostafa H, Maryam G, Hadi BF, Mahdi BF (2016) Effect of drought stress and foliar application of iron oxide nanoparticles on grain yield, ion content and photosynthetic pigments in sesame (*Sesamum indicum* L.). Iran J Field Crop Sci (Iran J Agric Sci) 46(4):619–628

Mukhopadhyay SS, Kaur N (2016) Nanotechnology in soil-plant system. Plant Nanotechnol 329–348

Mustafa G, Sakata K, Hossain Z, Komatsu S (2015) Proteomic study on the effects of silver nanoparticles on soybean under flooding stress. J Proteome 3:100–118

Nagajyoti PC, Lee KD, Sreekanth TVM (2010) Heavy metals, occurrence and toxicity for plants: a review. Environ Chem Lett 8(3):199–216

Nataliya VZ, Nataliya PD, Oksana ID (2014) Enhancement of drought resistance in wheat and corn by nanoparticles of natural mineral. Analcite Ecol Balkanica 6(1):1–10

Nezhadahmadi A, Prodhan ZH, Faruq G (2013) Drought tolerance in wheat. Sci World J 2013:610–721

Patra AK, Adhikari T, Bhardwaj AK (2016) Enhancing crop productivity in salt-affected environments by stimulating soil biological processes and remediation using nanotechnology. In: Innovative saline agriculture. Springer, New Delhi, pp 83–103

Praveen A, Khan E, Ngiimei D, Perwez M, Sardar M, Gupt M (2018) Iron oxide nanoparticles as nano-adsorbents: a possible way to reduce arsenic phytotoxicity in Indian mustard plant (*Brassica juncea* L). J Plant Growth Regul 37(2):612–624

Setter TL, Waters I (2003) Review of prospects for germplasm improvement for waterlogging tolerance in wheat, barley and oats. Plant Soil 253:1–34

Shabala S (2011) Physiological and cellular aspects of phytotoxicity tolerance in plants: the role of membrane transporters and implications for crop breeding for waterlogging tolerance. New Phytol 190:289–298

Sharon M, Choudhary AK, Kumar R (2010) Nanotechnology in agricultural diseases and food safety. J Phytol 2(4):83–92

Tamer E, Alaa EDO, Tarek A, Hassan E (2018) Nanomaterials and plant abiotic stress in agroecosystems. Environ Biodivers Soil Sec 2(1):73–94

Teakle NL, Real D, Colmer TD (2006) Growth and ion relations in response to combined salinity and waterlogging in the perennial forage legumes *Lotus corniculatus* and *Lotus tenuis*. Plant Soil 289:369–383

Tevini M (2004) Plant responses to ultraviolet radiation stress. In: Papageorgiou GC, Govindjee (eds) Chlorophyll a fluorescence. Advances in photosynthesis and respiration. Springer, Netherlands

Yaqoob S, Ullah F, Mehmood S, Mahmood T, Ullah M, Khattak A, Zeb MA (2018) Effect of waste water treated with TiO_2 nanoparticles on early seedling growth of *Zea mays* L. J Water Reuse Desalin 8(3):424–431

Ziska LH (2008) Rising atmospheric carbon dioxide and plant biology: the overlooked paradigm. In: Controversies in science and technology, from climate to chromosomes. Mary Ann Liebert, New Rochelle, NY

Chapter 4
Myco-nanotechnology in Agriculture

Khajista Jabeen and Faiza Anum

4.1 Introduction

A fungus is a eukaryotic organism that obtains its food by absorbing nutrients from the host through its cell wall. The branch of biology deals with fungus is known as mycology; mycology is originated from a Greek word μύκης *mykes* means mushroom. Fungi are extensively studied by Antonio de Michelli who is recognized as father of modern mycology. Fungal body (thallus) is composed of tubular microscopic thread like hyphae and reproduces by the formation of spores. Fungi differ from plants as they possess chitin in their cell wall instead of cellulose (Duran et al. 2005). Fungi are heterotrophic organisms which means they rely on other organisms or materials for their nutrition or food. They have not evolved the photosynthesis as a biological activity. They get nutrition by absorbing nutrients from their host (Parasites) or from dead organic material (decomposers) by secreting digestive enzymes extracellularly. Fungi are entitled to be the main decomposers of natural ecosystems. There is a great diversity in types of fungi and the habitat they live in. There is an estimation of 1,500,000 species of fungi on globe, but very few are identified. Now with the modern research tools, there is a hope that we may be successful to get more fungal strains in near future, with an approximation of nearly five million (Musarrat et al. 2010).

Fungi are considered as quite ordinary due to their smaller sizes and mysterious life cycles. Although they are having very important roles in ecosystems like they are parasites of plants and animals, symbiotic partners of algae and plants. Therefore they are having main and vital roles in recycling of nutrients between biotic and abiotic components of ecosystems (Heinrich and Wojewoda 1976). Some ecologically important and common types of fungi are given in Fig. 4.1.

K. Jabeen (✉) · F. Anum
Department of Botany, Lahore College for Women University, Lahore, Pakistan

S. Javad (ed.), *Nanoagronomy*, https://doi.org/10.1007/978-3-030-41275-3_4

Fig. 4.1 Some ecologically significant fungi (**a**) *Penicillium notatum* (**b**) *Lentinula edodes* (**c**) *Agaricus bisporus* (**d**) *Amanita muscaria*

Fungi reproduce asexually as well as sexually; sometimes a unique condition of reproduction also prevails in kingdom fungi known as parasexuallity. A fungus in which both asexual and sexual mode of reproduction are present is known as a Holomorphic fungus. If only asexual stage is present than fungus is said to be Anamorph while if only sexual stage is present than the fungus is a Telomorphic fungus.

There are different methods of asexual reproduction prevailing in the kingdom fungi. Most common of which are by spores, conidia and mycelial fragmentation. Spores are vegetative cells or bodies produced inside the sporangium while conidia are vegetative bodies cut off from the tips of certain hyphae known as conidiophores. Mycelial fragmentation is the phenomenon of breakdown of mycelia into small fragments and growth of each fragment into a new organism. Sexual reproduction gives two main benefits; one is the rapid and continuous dispersal of population and secondly to maintain the population adapted to a specific niche. Common modes of sexual reproduction in kingdom fungi are isogamy, anisogamy and oogamy. However, there is one class of fungi with no or unknown sexual stage in their life cycle and are known as imperfect fungi or Deuteromycota (Castro-Longoria et al. 2011). Fungi may have haploid, diploid or heterokaryon stage in their life cycle. In sexual life cycle of reproducing fungi, hyphae fuse to form a diploid hyphae or hyphae with heterokaryons (nuclei from the two parents don't fuse but stay together). This process is often referred as anastomosis and it is the first step for the sexual stage of the life cycle of such fungi (Shankar et al. 2003). Fungi are identified and grouped on the basis of morphology of fruiting bodies/sexual parts. For example members of Ascomycota and Basidiomycota have asci and basidia respectively as fruiting bodies containing ascospores and basidiospores.

4.2 Fungi and Agronomy

Fungi have both positive and negative impacts on agronomy. In one aspect fungi are notorious parasite and pathogen of large number of economically important crops. On the other hand fungi may act as biofertilizers in the form of mycorrhizal association with about 99% of the plant families. There are a number of pathogenic fungi causing different diseases in different plants. These diseases are mainly recognized and diagnosed by the type of host plant, plant part affected and the pattern/type of symptoms on infected part. A large number of fungi are parasitic and pathogenic in nature, as they use living host for their food, this makes them detrimental for crop plants (Kumamoto and Vinces 2005). Fungi parasitize various economically important plants including rice, wheat, potato, tomato, sugarcane, cotton. On basis of above facts, some of the general types of fungal diseases are discussed below

- Anthracnose
- Damping-off diseases
- Downy mildews
- Grey mold
- Leaf spots and blights
- Powdery mildews
- Root and foot rots
- Rusts
- Smuts

4.2.1 Damping-Off of Seedlings

Damping-off is a fungal disease in which fungus attack on growing seedlings. Damping-off is produced by various species of the fungal genera like *Pythium*, *Phytophthora*, *Rhizoctonia* and *Fusarium*. These all fungal strains are facultative parasites and cause almost same symptoms, so categorized as causative agents of damping off. The term "damping-off" is used because causative agents of this disease are mostly active in damp soils. As *Phytophthora* and *Pythium* species give rise to zoospores that need water for their movements in soil pores. Damping-off fungi are part of soil microbiota; they compete for organic material with other microbes present in the rhizosphere of the plants. If there is no organic material or host available for these fungi, they can form resistant resting structures like oospores in *Phytophthora* and *Pythium*, chlamydospores in *Fursarium solani* and sclerotia in *Rhizoctonia solani*.

This soil borne disease is categorized as important fungal disease causing much loss to the growing crops. It is characterized into two types

(a) Pre-emergent damping off which includes the decaying of seedlings and finally collapsing before they emerge from soil.
(b) Post emergent damping off which is the rotting and downfall of seedlings at soil level.

Seedlings of most plant species are susceptible to damping-off. This disease also attack on the fleshy and storage organs of important commercial plants. But when plants survived at the seedling stage, they are then not killed by damping-off fungi (Liao et al. 2000). Spores of damping off fungi (zoospores, oospores, mycelial hyphae) produce germ tubes and penetrate into seedlings tissues where they grow inter-cellularly as well as intra-cellularly (Kalo-Klein and Witkin 1990). They cause the breakdown of host tissue by secreting digestive enzymes and causing death to the plants.

4.2.2 Rot and Foot Rots

Root systems of a number of economically important crops are infected by "Rot and Foot Rots". These rots are caused by fungi and cause rotting of the roots and lower or basal portion of the stems adjacent to the roots. Due to the rotten roots, these plants can't absorb sufficient water and nutrients to maintain normal growth. As a result, stem and leaves are stunted and yellow; and they finally wilt and die. Sometimes damage due to infection is compensated or hidden by the growth of new roots (Volesky and Holan 1995).

There are four causative fungal groups for root rots. First group are the fungi with restricted host range and live as saprophytes, example includes *Gaeumannomyces graminis*. Second are the fungi which are saprophytic in nature and have a range of host crops that form resting spores e.g., *Rhizoctonia solani*. Third group includes *Sclerotinia sclerotiorum*, *Scterotium rolsfii*, which survive by producing sclerotia. Last group is host restricted but they do not form resistant bodies like *Fusarium* spp. such as *Fusarium oxysporum* and *Fusarium solani*.

4.2.3 Take-All of Wheat

It is also root disease found in wheat crop and caused by a fungus known as *Gaeumannomyces graminis* var. tritici. It is reported to be very common in areas where winters receive good rainfall and these soils are lighter, poorly drained, nutrient deficient with basic pH. This fungus may be restricted to a few plants in a field or may infect a larger area in patches.

Daniel McAlpine in 1902 described that this infection is also called "whiteheads" because of the destruction in root structure, growth of wheat plants is stunted and premature ripening occurs. Due to this earlier ripening, spikes appear as white heads among green normal spikes. These white head spikes are usually empty or with very small wrinkled grains. This may cause a serious damage to the crop and large reductions in the yield (Kumamoto and Vinces 2005). The take-all fungus survives in summer by infecting other herbs or grasses nearby in fields or by living saprophytically in the debris of wheat crop because it does not make resistant resting spores. This survival time period is highly influenced by the physical and chemical status of the soil environment as well as the local weather. It can survive for maximum of 2 years without host, therefore it is relatively easy to control.

4.2.4 Downy Mildews

The downy mildew fungi are obligate parasites and have a wide range of hosts. They cause severe losses to grains, vines, ornamentals and fruits. A number of fungal classes cause this infection. The name of disease is given due to the fact that primary infections are apparent on the surface of leaves, branches and fruits of infected plants in form of 'downy bloom'mainly containing sporangiophores and sporangia. Some of the common downy mildew causing fungi are *Plasmo parahalstedii*, *Sclerospora philippinsts*, *Plasmopora viticola* and *Perono sclerospora maydis*. These fungi cause systematic infections like at first chlorosis in form of streaks on leaves and then production of sporangia later on. Downy mildews can cause extensive damage to crops in years when the environment favors infection. A loss of US$250 million was estimated in USA and Canada due to the attack of downy mildew on tobacco crop (Liao et al. 2000).

4.2.5 Leaf Spot and Blight Diseases

This is common group of disease caused by a number of fungi and it's a very common infection in plant species, ranging from crops of agronomic importance to wild plants of natural communities. These infections can be seen on leaves and on other above ground parts of plants. The leaf spot causing fungi are usually belonging to the order Dothidiales of the Ascomycota group of fungi.

One of the famous example is Ascochyta blight of chickpea which caused a complete loss of crop a number of times. Disease is identified by the circular lesions on pods and foliage, elongated lesions on stem and petioles. It showed that whole aerial parts of plants come under attack of the pathogen. Circular rings like lesions on leaves and pods in fact have pycnidia, which produce two celled microspores in wet season. Therefore they come out in large numbers and attack tendrils of other plants. This disease prevails in areas of world with humid and cold climate and diseased seeds and crop leftovers are the main surviving spaces for the fungus between seasons (Mukherjee et al. 2001a).

4.2.6 Grey Mold Disease

Botrytis cinerea Pers. Ex Fr. is the causative agent of gray mold, one of the most destructive plant pathogen known to man. It causes both pre and post-harvest infections in a number of plants like strawberry, tomato, grapevine, chick pea, bulb flowers, cucumber, potato, onion and other ornamental plants. This disease is a serious concern in countries particularly where environment favors the growth of mold like Australia, Argentina, Pakistan, Nepal, India and Bangladesh (Shahiduzzaman 2015).

4.2.7 Rusts

Rusts are another diverse group, having some unique characteristics, causing havoc to crops of commercial importance. Causative agents belong to the order Uredinales of Basidiomycota and can infect and survive on living host only. Almost 168 rust genera and approximately 7000 species have been reported out of which half belongs to the genus Puccinia.

But luckily, steps can be taken to lower the losses caused by these pathogens as they are very much host specific but don't cause the infection in non-host plants. A hurdle in research is the resistance in growth in pure cultures by these fungi, which causes delays in the research process at lab level. Rust fungi complete their life cycles in two separate hosts with different types of spores. These different types of spores are host specific in fact and cannot grow on the other host. Infections of rust fungi are marked by the rust like fruiting bodies having spores on the surface of leaves, petioles, stem, tender shoots, fruits etc. These rust spots may be of different colors like yellow, orange, black etc. Rust infection causes the stunted appearance of the infected plants with yellow leaves (Latif et al. 2018).

4.2.8 *Smuts*

The word "Smut" is derived from the German word which means smoke or dirt. This name was given to the infection because the inflorescence infected by smut releases the fungal spores in the form of smoke or dust. Smuts are pathogens of cereal crops infecting the members of Poacea and Cyperaceae. Important host plants are wheat, maize, rye, grass, oat, sugarcane and other grasses. Smuts belonging to the class Ustilaginales of class Basiodiomycota, mainly attack on leaves and stems by developing sori within the plant tissue. Finally they hijack the inflorescence and reproductive parts of the plant where they form galls full of thick walled black teliospores which when released gives smoky appearance and hence get the name (Wunderle et al. 2012).

4.2.9 *Anthracnose*

Anthracnose is caused by *Colletotrichum gloeosporioides* and characterized as appearance of dark lesions. It has been reported to cause yield losses (34–47%) to fruit plants especially mango. This disease affects a number of plants but most affected crop is of mango. It infects the flower sets of mango trees and caused the losses in fruit production in warm and humid conditions of climate (Sundravadana et al. 2007; Pandey et al. 2012). *C. gloeosporioides* secretes various enzymes like Polygalacturonase, Polygalacturonase transeliminase and cellulase that might be responsible for its pathogenicity (Jat et al. 2017).

In conventional agricultural practices, fungi are mainly controlled by use of different chemicals which are also called fungicides. But there is an increasing public concern about the overuse and misuse of fungicides. These fungicides on one side control the fungal diseases but on the other hand environmental pollution is cumulative day by day due to over use of fungicide. Increased research in the area of fungal role in agronomy can make a milestone in modern agronomic techniques. An alternative of fungicide is the use of natural compounds or biological control. In nature many plant and plant families possess antifungal constituents in the form of secondary metabolites. Use of nanotechnology and nanoparticles is another solution to avoid fungicide consumption to protect environment.

4.3 Role of Fungi in Nanoparticles Synthesis

There are large numbers of living organisms, which give a chance to biotechnologists to explore and exploit them for the wellbeing of human race. One of them is to synthesize nanoparticles by using fungal organisms directly or by using their metabolites and extracts. Huge amounts of fungal enzymes which take part in NPs

Table 4.1 List of fungi that synthesize metal nanoparticles

Sr. #	Type of nanoparticles	Fungal species	References
1	Ag (silver nanoparticles)	*Alternaria alternata*	Kareem et al. (2019), Sarkar et al. (2011)
		Alternaria solani	Amal and Ghazwani (2015)
		Aspergillus flavus	Ninganagouda et al. (2013)
		Aspergillus niger	Gaikwad and Bhosale (2012)
		Fusarium oxyporum	Amal and Ghazwani (2015)
2	Au (gold nanoparticles)	*Alternaria alternata*	Dhanasekar et al. (2015), Sarkar et al. (2012)
		Aspergillus sidowii	Vala (2015)
		Rhizopus oryzae	Das et al. (2009)
		Thermophillic fungi	Molnar et al. (2018)
		Rhodococcus sp	Ahmad et al. (2003)
3	Cu (copper nanoparticles)	*Neurospora crassa*	Rashmi et al. (2004)
		Stereum hirsutum	Cuevas et al. (2015)
		Ureolytic fungi	Li and Gadd (2017)
4	Fe (iron nanoparticles)	*Alternaria alternata*	Mohamed et al. (2015)
		Aspergillus oryzae	Raliya (2013)
		Verticillium sp.	Bharde et al. (2006)
5	Si (silica nanoparticles)	fungus	Bansal et al. (2005)
6	Zn (zinc nanoparticles)	*Aspergillus fumigatus*	Velmurugan et al. (2010) Baskar et al. (2013)

formation can be obtained on commercial level in fermenters (Pimprikar et al. 2009). Myconanotechnology is a new field of research making its roots in last decade. To synthesize metal nanoparticles with success, a substantial number of fungal species are widely used like *Fusarium oxysporum*, *Rhizopus oryzae* and *Verticillium* sp. (Table 4.1). The utilization of biomass of fungi and/or cell free extract for the synthesis of metal NPs yielded in different shapes and sizes of these myconanoparticles (Narayanan and Sakthivel 2010). Both spherical and quasi-spherical silver NPs were produced from *Fusarium oxysporum* with size varying from 20 to 50 nm (Bharde et al. 2006). Extract of *Rhizopus oryzae* was used in manufacturing of the diverse shape gold nanoparticles and it depends on gold particle concentration, pH value and reaction time (Binupriya et al. 2010).

4.3.1 Mechanisms Behind Myconanoparticles Synthesis

Several promising mechanisms have been recommended for the development of metal nanoparticles, but no such mechanism has been known yet and extensive research is still needed. According to Mukherjee et al. (2001a, b) mainly the cell wall and sugar component of the fungal cell wall involve in the process of bio-reduction of the metallic ions. Nanoparticles are formed on the exterior of fungal cell wall, and the very basic step is bio-reduction to trap the metallic ions. The electrostatic interactions between charged group on the cell wall surface and metal ions followed by metal ions enzymatic reduction lead to the accumulation and formation of nanoparticles.

Birla et al. in 2009 suggested that fungal cell wall proteins also play a substantial role in the formation of zirconia nanoparticles. Fungi secrete hydrolyzing protein in acidic condition that makes binding with zirconium to form the zirconium NPs. These NPs forming proteins are cationic in nature with a molecular weight of 55 kDa. *Verticillium* sp. also produces these cationic proteins which might be the cause of hydrolysis of ferric ions (Bharde et al. 2006). According to Ahmad et al. (2003) tryptophan and tyrosine are the amino acids that play a pivotal role in the bio reduction of metal ions to metallic nanoparticles. NADH-dependent enzymes and the fungal proteins are also involved in the metal ion reduction (Germain et al. 2003).

4.3.2 Fungi a Renewable Source for Nanoparticles Synthesis

Fungi are fascinating source for the green synthesis of nanoparticles owed to their metal bioaccumulation capacity. Furthermore, fungi are easy to grow in the laboratory and production of large quantity of biomass make them valuable to be used in the green synthesis of NPs (Kumar et al. 2011). Fungal cell wall possess a number of functional groups that make bonds with the metal ions and this high wall-metal binding capacity makes fungi more attractive than other microbes (Maynard and Michelson 2006). Biosynthesized nanoparticles are environment friendly and are biocompatible for pharmacological uses. Biosynthesis of nanoparticles by fungi as a base material may be a reasonable approach. Fungal enzymes possess high redox potential which makes them more suitable for the oxidation reduction reaction for the conversion of metallic ions into specific nanoparticles. So, the green synthesis of nanoparticles is now an attractive field around the globe. Figure 4.2, Illustrates Advantages of fungi as bio-factories for nanoparticles synthesis.

4.4 Myconanoparticles Application in Management of Fungal Diseases

Myconanotechnology is a rising field, in which fungi can control the synthesis of nanostructures with required size and form. Mycosynthesis of various types of nanoparticles is carried out using metal salts such as silver, gold, titanium, selenium, platinum, palladium, zirconia, tellurium, cadmium, telluride, silica and magnetite gold-silver alloy. Currently, silver nanoparticles have attracted the scientists due to their extensive application in the areas like biomedicine, agriculture, physics and microbial biotechnology as antimicrobial agents at the nano-scale (Kowshik et al. 2002).

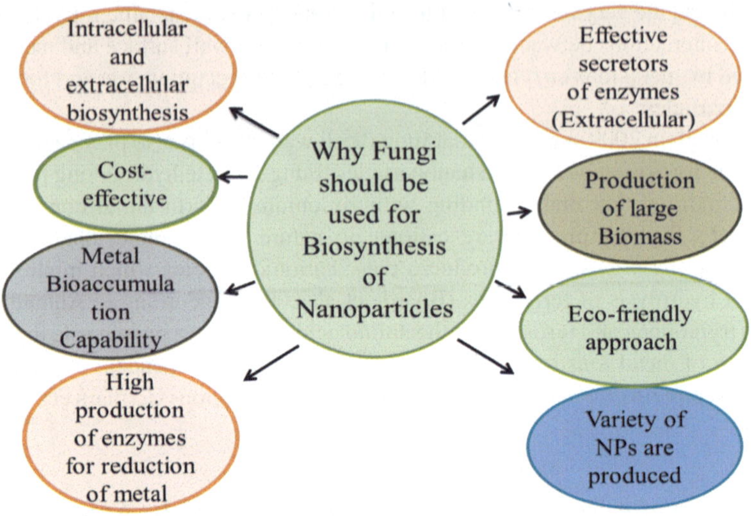

Fig. 4.2 Advantages of fungi as bio-factories for nanoparticles synthesis

Phytopathologists are working in search of techniques to protect economically important crops from destructive plant pathogens. Nanotechnology provides as alternative way for the production of pesticides and fertilizers which are safer to environment. Sanghi and Verma (2010) suggested that nanoparticles are found effective against pests, nematodes and fungal plant pathogens.

Though, fungus based green synthesis of myconanoparticles and their extensive range of applications (Fig. 4.3) have fascinated the attention of investigators. Several noticeable applications of nanotechnology in different areas of agronomy have been described in subsequent sections:

4.4.1 Nanoparticles as a Suppresser for Pests (Nanopesticides)

The use of AgNPs as antimicrobial agents has been exploited in controlling plant diseases. Nanoparticles have potentially shown various modes of inhibitory action against plant pathogens. The use of biosynthesized AgNPs by exploiting biological agents for the effective management of plant disease has attained significant consideration now a day. The reason of this approach is the low cost and less toxicity of nanoparticles to biomolecules. Furthermore, biosynthesized AgNPs are equally capable to those nanoparticles produced by chemical methods. Nanoparticles offer non-toxic environment friendly method for the management of plant diseases against phytopathogens (Feng et al. 2000).

Plants are directly or indirectly influenced by various types of biotic stress such as disease and as well as abiotic stress such as drought, salinity, heat, flood, etc.

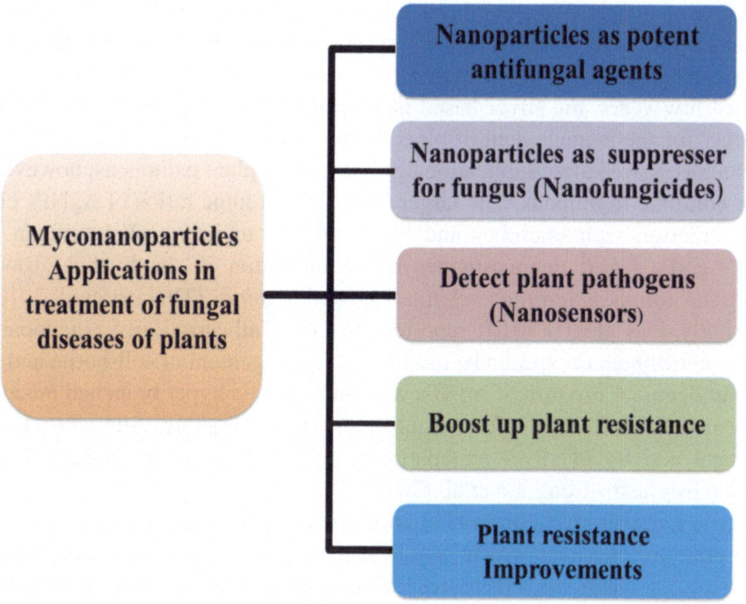

Fig. 4.3 Potential Myconanoparticles applications in plant pathology

which is responsible for destructive economic loss. Nowadays, to avoid such type of damages, breeders developed resistant varieties of plants against them. The Nano-biotechnology improves plant resistance by implying novel measures using NPs like, nano capsules and nanofibers, to manipulate genes to increase plant resistance (Mukherjee et al. 2002).

It is possible to use nanoparticles against various types of plant pathogens because it is safer in comparison to synthetic fungicides, for example-Ag-SiO_2 NPs have significant antifungal activity against *Botrytis cinerea* (Riddin et al. 2006). Germain et al. (2003) also reported that combined application of a fungicide such as fluconazole and Ag NPs, can effectively suppress the growth of *Phomaglomerata, Phomaherbarum, Candida albicans Trichoderma* sp. and *Fusarium semitectum.* Soil borne plant pathogens produce hard structures in the form of sclerotia which are difficult to manage. These sclerotia are basically survival structures contain melanin to overcome environmental stress and are resistant to chemical and physical degradation.

Sclerotium rolfsii is a devastating soil borne plant pathogen which causes collar rot in chickpea and leads to about 50–95% mortality at seedling stage (Gericke and Pinches 2006). Sun and Xia (2002) suggested that there is a direct interface between biosynthesized AgNPs and sclerotia. The scleortia of *S. rolsfii* show reduced germination by the application of AgNPs as these AgNPs produce very responsive Ag^+ ions which then show their antimicrobial activity to stop the growth of fungus. Unquestionably the diffusion into the microbial cell, disturbance in transport system, accretion of Ag^+ ions and fabrication of reactive oxygen species are the certain approaches of movements of AgNPs responsible for its antimicrobial property (Sarkar et al. 2012).

In last few years, the silver based nanopesticides are rapidly developed by the researcher for the management of plant pathogens due to their antimicrobial property which is proved after application against various plant pathogens; however, it is safe or nontoxic to humans. The larger surface-to-volume ratios of AgNPs elevate their interaction with microbes and their capacity to infuse (Velmurugan et al. 2010). AgNPs based nanopesticides cause inhibition in the hyphal growth of *Rhizoctonia solani*, *Sclerotium sclerotiorum* and *S. minor* (Thakkar et al. 2010).

Similarly Das et al. (2010) reported that biosynthesized silver nanoparticles using *Stenotrophomoas* sp. can be used for the management of soil-borne and foliar phytopathogens. A mixture of AgNPs with amphiphilic hyper branched macromolecules can be utilized as an effective antifungal surface coating (Ahmad et al. 2006). Antifungal activity of silver nanoparticles against 18 plant pathogens was also described by Castro-Longoria et al. (2011).

Silica is known to have prominent antifungal potential as it increases the hydrophobic pressure of leaves which may trigger the plant defense towards biotic and abiotic stresses. Nano-silica increases the defense response towards fungal infection in maize over bulk silica treatment (Verma et al. 2010). This is possibly because of the mechanism of increased silica transport and deposition in roots as well as in leaves. Thus, it is evident that nanoparticles or nanoparticle aggregates with diameter less than the pore diameter of the cell wall can easily penetrate and reach plasma membrane. There is an enlargement of pores or induction of new cell wall pores on interaction with nanoparticles which in turn enhances the uptake of water and nutrients of plants (Tian et al. 2010). In addition, deposition of amorphous silica in the cell walls leads to leaf erectness and hence, may prevent the invasion of pathogenic fungi. Thus the nanosilica can be used for active defense mechanism to improve crop protection.

4.4.2 Myconanoparticles Mechanism of Action Against Fungal Pathogens

The exact mechanism of growth arrest and destruction of fungal pathogen of plants by myconanoparticles is not well understood. Microorganisms are believed to use some enzymes to metabolize oxygen to sustain life. Silver ion particles scrapple the enzyme and stop oxygen metabolization. This suffocates the fungi and other microorganism, leading to death (Longoria et al. 2012). Currently, fungus based AgNPs are being completely surveyed and extensively investigated as potential antifungal agents. Their tiny size and high surface-to-volume ratio boost their interaction with fungal species to carry a diverse range of antifungal activities (Blackwell 2011). The suppressive effect of AgNPs on fungi has led to follow many mechanisms. It is assumed that AgNPs possess high affinity towards sulphur and phosphorus in the cell. As silver ions (Ag^+) from AgNPs move with DNA containing phosphorous element moieties, this leads to inactivation of DNA replication. Interaction of AgNPs with proteins containing sulphur which is present within or outside the cell is considered to be another reason. AgNPs become connected to sulphur-containing proteins of microorganism cell membranes resulting in high permeability of cell membrane, leading to the death of microorganisms (Ahmad et al. 2002). Studies on the dose dependent effects of AgNPs (in the range of 10–15 nm) on microorganisms showed that at a micromolar level of these Ag^+ ions can inhibit enzymes participating in respiration or may interfere in permeability of membrane to protons and phosphate. At higher concentrations these ions can move within nucleic acids and cytoplasmatic elements (Das et al. 2009). As fungi contain large quantities of protein product, they're most popular as new economical sources for nanostructure formulation (Alvarez-Puebla et al. 2004).

4.5 Future Prospective

The production of metallic nanoparticles by fungi appears to be a relatively simple biotechnological process, predominantly involving only the reaction of fungal culture filtrates with solutions of metal salts. There are, however, a number of issues where further research is required. The work undertaken to date indicates that there may be a number of different kinds of reducing agents involved in the mechanism of synthesis of metal nanoparticles. These may also have effects on the final shapes and size of the nanoparticles. There is also a need to search out the specific mechanisms which involved in nanoparticle formation from fungi, and for comparative studies to determine whether the same or different pathways are used by different fungi for different metals. Once the basic mechanisms have been determined there will be a need to evolve mechanism for optimizing the specific concentrations, sizes and shapes of the nanoparticles. For the synthesis of commercially feasible myconanaoparticles, it would be necessary to develop low-cost recovery techniques to isolate the nanoparticles from the fungal mycelium for the easy use of these NPs in industrial processes. More researches need to be carried out on applications of myco-nanoparticles for plant disease treatment by a targeting particular disease and their controlled release.

References

Ahmad A, Mukherjee P, Mandal D, Senapati S, Khan MI, Kumar R, Sastry M (2002) Enzyme mediated extracellular synthesis of CdS nanoparticles by the fungus, *Fusarium oxysporum*. J Am Chem Soc 124:12108–12109

Ahmad A, Senapati S, Khan MI, Kumar R, Ramani R, Srinivas V, Sastry M (2003) Intracellular synthesis of gold nanoparticles by a novel alkalotolerant actinomycete, *Rhodococcus* species. Nanotechnology 14(7):824–828

Ahmad Z, Pandey R, Sharma S, Khuller GK (2006) Alginate nanoparticles as antituberculosis drug carriers: formulation development, pharmacokinetics and therapeutic potential. Ind J Chest Dis Allied Sci 48(3):171
Alvarez-Puebla RA, Dos Santos JDS, Aroca RF (2004) Surface-enhanced Raman scattering for ultrasensitive chemical analysis of 1 and 2-naphthalenethiols. Analyst 129(12):1251–1256
Amal A, Ghazwani AA (2015) Biosynthesis of silver nanoparticles by *Aspergillus niger, Fusarium oxysporum* and *Alternaria solani*. Afr J Biotechnol 14(26):2170–2174
Bansal V, Rautaray D, Bharde A, Ahire K, Sanyal A, Ahmad A, Sastry M (2005) Fungus-mediated biosynthesis of silica and titania particles. J Mater Chem 26:2583–2589
Baskar G, Chandhuru J, Sheraz Fahad K, Praveen AS (2013) Mycological synthesis, characterization and antifungal activity of zinc oxide nanoparticles. Asian J Pharm Technol 3(4):142–146
Bharde A, Rautaray D, Bansal V, Ahmad A, Sarkar I, Yusuf SM, Sastry M (2006) Extracellular biosynthesis of magnetite using fungi. Small 2(1):135–141
Binupriya AR, Sathishkumar M, Yun SI (2010) Biocrystallization of silver and gold ions by inactive cell filtrate of *Rhizopus stolonifera*. Colloids Surf B: Biointerfaces 79(2):531–534
Birla SS, Tiwari VV, Gade AK, Ingle AP, Yadav AP, Rai MK (2009) Fabrication of silver nanoparticles by *Phoma glomerata* and its combined effect against *Escherichia coli, Pseudomonas aeruginosa* and *Staphylococcus aureus*. Lett Appl Microbiol 48(2):173–179
Blackwell M (2011) The Fungi: 1, 2, 3… 5 1 million species? Am J Bot 98(3):426–438
Castro-Longoria E, Vilchis-Nestor AR, Avalos-Borja M (2011) Biosynthesis of silver, gold and bimetallic nanoparticles using the filamentous fungus *Neurospora crassa*. Colloids Surf B: Biointerfaces 83(1):42–48
Cuevas R, Durán N, Diez MC, Tortella GR, Rubilar O (2015) Extracellular biosynthesis of copper and copper oxide nanoparticles by *Stereum hirsutum*, a native white-rot fungus from Chilean forests. J Nanomater Arch 16(1):1–7. https://doi.org/10.1155/2015/789089
Das SK, Das AR, Guha AK (2009) Gold nanoparticles: microbial synthesis and application in water hygiene management. Langmuir 25(14):8192–8199
Das SK, Das AR, Guha AK (2010) Microbial synthesis of multishaped gold nanostructures. Small 6(9):1012–1021
Dhanasekar NN, Rahul GR, Narayanan KB, Raman G, Sakthivel N (2015) Green chemistry approach for the synthesis of gold nanoparticles using the fungus *Alternaria* sp. J Microbiol Biotechnol 25(7):1129–1135
Duran N, Marcato PD, Alves OL, De Souza GI, Esposito E (2005) Mechanistic aspects of biosynthesis of silver nanoparticles by several *Fusarium oxysporum* strains. J Nanobiotechnol 3(1):8
Feng QL, Wu J, Chen GQ, Cui FZ, Kim TN, Kim JO (2000) A mechanistic study of the antibacterial effect of silver ions on *Escherichia coli* and *Staphylococcus aureus*. J Biomed Mater Res 52(4):662–668
Gaikwad S, Bhosale A (2012) Green synthesis of silver nanoparticles using *Aspergillus niger* and its efficacy against human pathogens. Eur J Exp Biol 2(5):1654–1658
Gericke M, Pinches A (2006) Biological synthesis of metal nanoparticles. Hydrometallurgy 83(1–4):132–140
Germain V, Li J, Ingert D, Wang ZL, Pileni MP (2003) Stacking faults in formation of silver nanodisks. J Phys Chem 107(34):8717–8720
Heinrich Z, Wojewoda W (1976) The effect of fertilization on a pine forest ecosystem in an industrial region IV Macromycetes. Ekologia Polska 24:319–330
Jat RD, Nanwal RK, Jat HS, Bishnoi DK, Dadarwal RS, Kakraliya SK et al (2017) Effect of conservation agriculture and precision nutrient management on soil properties and carbon sustainability index under maize–wheat cropping sequence. Int J Chem Stud 5(5):1746–1756
Kalo-Klein A, Witkin SS (1990) Prostaglandin E_2 enhances and gamma interferon inhibits germ tube formation in *Candida albicans*. Infect Immun 58:260–262
Kareem SO, Familola OT, Oloyede AR, Dare EO (2019) Microbial synthesis of silver nanoparticles using *Alternaria alternata* and their characterization. Appl Environ Res 41(1):1–7

Kowshik M, Ashtaputre S, Kharrazi S, Vogel W, Urban J, Kulkarni SK, Paknikar KM (2002) Extracellular synthesis of silver nanoparticles by a silver-tolerant yeast strain MKY3. Nanotechnology 14(1):95–98

Kumamoto CA, Vinces MD (2005) Alternative *Candida albicans* lifestyles: growth on surfaces. Annu Rev Microbiol 59:113–133

Kumar D, Karthik L, Kumar G, Roa KB (2011) Biosynthesis of silver nanoparticles from marine yeast and their antimicrobial activity against multidrug resistant pathogens. Pharmacology Online 3:1100–1111

Latif M, Hassan T, Shad GM, Ahmad G, Sajjid AR, Nawaz A, Ahmad M (2018) Comparison of rust infection with area on different varieties of wheat in district Sialkot. Int J Adv Multidiscip Res 5(2):1–6

Li Q, Gadd GM (2017) Biosynthesis of copper carbonate nanoparticles by ureolytic fungi. Appl Microbiol Biotechnol 101(19):7397–7407

Liao DI, Basarab GS, Gatenby AA, Jordan DB (2000) Selection of a potent inhibitor of trihydroxynaphthalene reductase by sorting disease control data. Bioorg Med Chem Lett 10(5):491–494

Longoria EC, Velasquez SM, Nestor AV, Berumen EA, Borja MA (2012) Production of platinum nanoparticles and nanoaggregates using *Neurospora crassa*. J Microbiol Biotechnol 22(7):1000–1004

Maynard A, Michelson E (2006) The nanotechnology consumer products inventory. Woodrow Wilson International Center for Scholars, Washington, DC. Accessed 23 Mar

Mohamed YM, Azzam AM, Amin BH, Safwat NA (2015) Mycosynthesis of iron nanoparticles by *Alternaria alternata* and its antibacterial activity. Afr J Biotechnol 14(14):1234–1241

Molnar Z, Bódai V, Szakacs G, Erdélyi B, Fogarassy Z, Sáfrán G, Varga T, Kónya Z, Tóth-Szeles E, Szűcs R, Lagzi I (2018) Green synthesis of gold nanoparticles by thermophilic filamentous fungi. Sci Rep 8(3943):1–10

Mukherjee P, Ahmad A, Mandal D, Senapati S, Sainkar SR, Khan MI, Parishcha R, Ajaykumar PV, Alam M, Kumar R, Sastry M (2001a) Fungus-mediated synthesis of silver nanoparticles and their immobilization in the mycelial matrix: a novel biological approach to nanoparticle synthesis. Nano Lett 11:515–519

Mukherjee P, Ahmad A, Mandal D, Senapati S, Sainkar SR, Khan MI, Sastry M (2001b) Bioreduction of $AuCl_4^-$ ions by the fungus, *Verticillium* sp and surface trapping of the gold nanoparticles formed. Angew Chem 40(19):3585–3588

Mukherjee P, Senapati S, Mandal D, Ahmad A, Khan MI, Kumar R, Sastry M (2002) Extracellular synthesis of gold nanoparticles by the fungus *Fusarium oxysporum*. Chembiochem 3(5):461–463

Musarrat J, Dwivedi S, Singh BR, Al-Khedhairy AA, Azam A, Naqvi A (2010) Production of antimicrobial silver nanoparticles in water extracts of the fungus *Amylomyces rouxii* strain KSU-09. Bioresour Technol 101(22):8772–8776

Narayanan KB, Sakthivel N (2010) Biological synthesis of metal nanoparticles by microbes. Adv Colloid Interface Sci 156(1–2):1–13

Ninganagouda S, Rathod V, Jyoti H, Singh D, Prema K, Haq M (2013) Extracellular biosynthesis of silver nanoparticles using *Aspergillus flavus* and their antimicrobial activity against gram negative MDR strains. Int J Pharm Biol Sci 4(2):222–229

Pandey A, Yoon H, Lyver ER, Dancis A, Pain D (2012) Identification of a Nfs1p-bound persulfide intermediate in Fe–S cluster synthesis by intact mitochondria. Mitochondrion 12(5):539–549

Pimprikar PS, Joshi SS, Kumar AR, Zinjarde SS, Kulkarni SK (2009) Influence of biomass and gold salt concentration on nanoparticle synthesis by the tropical marine yeast *Yarrowia lipolytica* NCIM 3589. Colloids Surf B Biointerfaces 74(1):309–316

Raliya R (2013) Rapid, low-cost, and ecofriendly approach for iron nanoparticle synthesis using *Aspergillus oryzae* TFR9. J Nanomater 2013:1–3. https://doi.org/10.1155/2013/530170

Rashmi K, Krishnaveni T, Ramanamurthy S, Mohan PM (2004) Characterization of cobalt nanoparticle from a cobalt resistant strain of *Neurospora crassa*. In: International symposium of research students on materials science and engineering, Dec 2004, Chennai

Riddin TL, Gericke M, Whiteley CG (2006) Analysis of the inter- and extracellular formation of platinum nanoparticles by *Fusarium oxysporum* sp lycopersici using response surface methodology. Nanotechnology 17(14):3482–3489

Sanghi R, Verma P (2010) pH dependent fungal proteins in the 'green' synthesis of gold nanoparticles. Adv Mater Lett 1(3):193–199

Sarkar J, Chattopadhyay D, Patra S, Deo SS, Sin S (2011) *Alternaria alternata* mediated synthesis of protein capped silver nanoparticles and their genotoxic activity. Digest J Nanomater Biostruct 6(2):563–573

Sarkar J, Ray S, Chattopadhyay D, Laskar A, Acharya K (2012) Mycogenesis of gold nanoparticles using a phytopathogen *Alternaria alternata*. Bioprocess Biosyst Eng 35(4):637–643

Shahiduzzaman M (2015) Efficacy of fungicides in controlling *Botrytis* grey mold of chick pea (*Cicer arietinum* L.). Bangladesh J Agric Res 40(3):391–398

Shankar SS, Ahmad A, Pasricha R, Sastry M (2003) Bioreduction of chloroaurate ions by geranium leaves and its endophytic fungus yields gold nanoparticles of different shapes. J Mat er Chem 13(7):1822–1826

Sun Y, Xia Y (2002) Shape-controlled synthesis of gold and silver nanoparticles. Science 298(5601):2176–2179

Sundravadana S, Alice D, Kuttalam S, Samiyappan R (2007) Efficacy of azoxystrobin on *Colletotrichum gloeosporioides* Penz growth and on controlling mango anthracnose. J Agric Biol Sci 2(3):10–15

Thakkar KN, Mhatre SS, Parikh RY (2010) Biological synthesis of metallic nanoparticles. Nanomed Nanotechnol Biol Med 6(2):257–262

Tian X, He W, Cui J, Zhang X, Zhou W, Yan S, Yue Y (2010) Mesoporous zirconium phosphate from yeast biotemplate. J Colloid Interface Sci 343(1):344–349

Vala KA (2015) Exploration on green synthesis of gold nanoparticles by a marine-derived fungus *Aspergillus sydowii*. Environ Prog Sustain Energy 34(1):194–198

Velmurugan P, Shim J, You Y, Choi S, Kamala-Kannan S, Lee KJ, Oh BT (2010) Removal of zinc by live, dead, and dried biomass of *Fusarium* spp isolated from the abandoned-metal mine in South Korea and its perspective of producing nanocrystals. J Hazard Mater 182(1–3):317–324

Verma VC, Kharwar RN, Gange AC (2010) Biosynthesis of antimicrobial silver nanoparticles by the endophytic fungus *Aspergillus clavatus*. Nanomedicine 5(1):33–40

Volesky B, Holan ZR (1995) Biosorption of heavy metals. Biotechnol Prog 11(3):235–250

Wunderle J, Leclerque A, Schaffrath U, Slusarenko A, Koch E (2012) Assessment of the loose smut fungi (*Ustilago nuda* and *U. tritici*) in tissues of barley and wheat by fluorescence microscopy and real-time PCR. Eur J Plant Pathol 133(4):865–875

Chapter 5
Nanotechnology in Pest Management

Iqra Akhtar, Zunera Iqbal, and Zeb Saddiqe

5.1 What Is the Science of Nanotechnology?

Nanotechnology has been emerged in different fields of applied sciences like, biology, medicine, chemistry, engineering and physics. The "formation, characterization and application of materials, system and devices by regulating the shape and sizes at the nanoscale" is known as Nanotechnology (Ramsden 2005).

A meter comprises of one billion of nanometers, so 1 nm is one billionth part of a meter. In general word Nano indicates the size range from one nanometer (1 nm) to hundred nanometer (100 nm) in one dimension. It requires the development and modification of devices and materials between that size ranges. Figure 5.1 describes different ranges and types of nanoparticles.

5.2 Pest and Pesticides

The unwanted organisms which are harmful and cause damage to humans, animals and crops are called "Pests". The chemical substances which are used for killing, repelling, preventing, mitigating and destroying the pests are known as "Pesticides".

There are different groups of pests which cause different kinds of damage to humans, animals and crops. To identify the type of damage caused by pests, would helpful to understand and identify the type of pest. There are five main groups in which pests can be divided:

1. **Arthropods**: It includes Spiders, Insects, Ticks and Mites.
2. **Nematodes**: It includes Root knot and Root lesion nematodes.
3. **Vertebrates**: This group contains animals with backbones.

I. Akhtar (✉) · Z. Iqbal · Z. Saddiqe
Department of Botany, Lahore College for Women University, Lahore, Pakistan

S. Javad (ed.), *Nanoagronomy*, https://doi.org/10.1007/978-3-030-41275-3_5

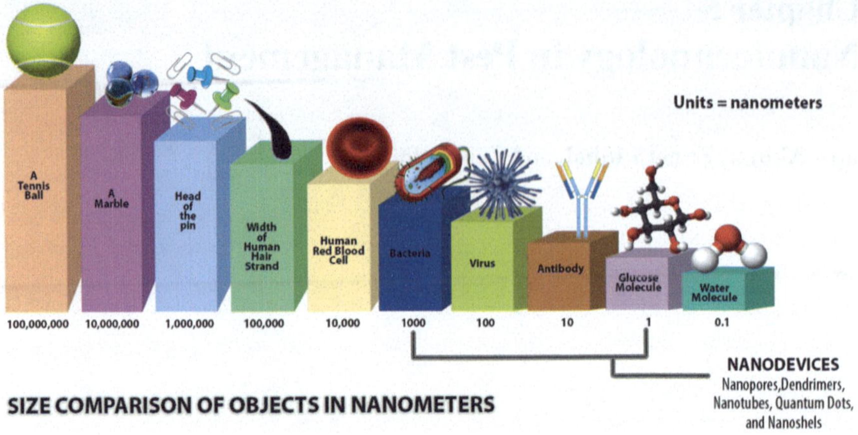

Fig. 5.1 Different ranges and types of nanoparticles (Brar et al. 2010)

4. **Weeds**: This group contains dandelion, nutsedge, pampas grass and purslane.
5. **Pathogens**: It contains the microorganisms which cause diseases.

There are many types of pesticides which are commonly used in the field of agriculture. These pesticides includes Azoxystrobin (used on vegetables, cereals, fruits and rice), Atrazine (used to eliminate harmful weeds from major crops), Metolachlor (used to control weeds in corn and cotton), Pendimethalin (used to control pest in soyabean, wheat and annual grasses), and Oxyfluorfen (used to control weeds in fruits and vegetables).

5.2.1 Risk Factors Associated with Pesticides

Due to the poisonous properties of pesticides there are many risk factors which are correlated with the use of pesticides such as:

The users of the pesticides:

One of the main risk associated with the use of pesticides is the farmers and their family members. They are at the highest risk. When the farmers mix the drugs or chemicals and when they apply these chemicals to the crops are examples of their direct contact with these pesticides.

The consumers of farm products:

When the pesticides are applied to the crops, these chemicals leaves residues behind and those residues will be used by the consumers.

The environment:

Pesticides not only kills the targeted organisms but also kills the other beneficial organisms including fish, birds, earthworms and insects which are present in or all around the fields, which cause the loss of biodiversity, results in the death of farm animals and death of wildlife. These poisonous chemicals can easily get mixed with water bodies, air and soil to contaminate it.

5.3 Pest and Agriculture Today

Insecticides, herbicides, nematicides, fungicides, plant growth regulators, rodenticides and molluscicides are all included in the term pesticides as it covers a large range of compounds. Generally a pesticide must have the properties like destructive to the specified pests and non-harmful to non-targeted species, like human. Unfortunately, these properties are not considered while using the pesticides so that the arguments about use and misuse of pesticides has surfaced. The excessive use of these pesticides, as considering the proverb, "If little is good, a lot more will be better" has played destructive role in mankind and other forms of life (Aktar et al. 2009).

5.3.1 Excessive Use of Pesticides

Crop productivity is limited by the major factor known as "diseases". The problem is that the disease management could not identify and detect the exact situation and stage of disease prevention. Pesticides are usually applied by no consideration of precautionary measures and their harmful aspects, as a result, residual toxicity and environmental threats come into fronts. But on the other side, when pesticides are applied after the appearance of the disease, it leads to the loss of crops to some extent. Diseases caused by viruses (viral diseases) are very difficult to control in comparison to all other types of diseases, because in viral diseases the spreading of disease takes place by the vector and it is very difficult to stop the spreading. As at the initial stages, it starts to show the symptoms of the disease then the use of pesticides would not be more effective. That's why the identification and detection of the specific stage of viral disease such as DNA replication and production of viral proteins is the main point to successfully control the viral disease.

To improve, the quality of food, taste and better flavor, techniques of biotechnology are well known. It has been very safe to use biotechnological techniques in agriculture and to increase the crop preservation, production and protection. As biotechnology has many benefits but at the same time it has many ecological consequences which are harmful for the environment. These harmful impacts include, wide spreading of genetically modified genes to the native plants which results in increasing the toxicity level. This toxicity level may also pass through the food chain and can disturb the natural system of pest control by generating the new viral resistant strains and weeds as a result of loss of biodiversity and occurrence of insecticidal resistance. Therefore, it is really important to bring new, modern and inventive methods and technologies to sort out the above mentioned problems (Kitherian 2017).

A new area of biotechnological research included the detection, identification and application of biomarkers. These biomarkers point out the exact stage of disease. The identification and detection of other proteins produced during the period of infection can be carried out by comparing the protein production in diseased state with the protein production in healthy state.

Rats, mice, ticks, mosquitoes, other disease carriers and pests are controlled by using pesticides. In agriculture to control disease, weeds and insect infection, pesticides are used. Nowadays the use of pesticides is the fastest way to overcome the effects of different disease and pests. These chemicals include herbicides, fungicides and pesticides. Biological control methods are also used but presently these are very expensive. Problems associated with uncontrolled use of pesticides includes:

- Deleterious effects on the human health.
- Damaging effects on domestic animals and pollinating insects.
- Penetration of these materials into the soil and getting mixed with the water.
- Having direct and indirect effects on ecosystem.

5.4 Nanotechnology for Pest Management and Disease Control

The appropriate and suitable solution for the above mentioned problems is the controlled and programmed use of chemicals on nano scale basis. In controlled release method these materials are directly used into the plant's part where the pest or disease attacked. These carriers are self-regulatory on nano scale, which means that only the required amount of chemical is delivered to the plant tissue as medication. Such nano particles are also used to change the kinetic profile of the releasing drugs which results in more appropriate and sustainable release of drugs with lesser demand for constant dosing (Sharon et al. 2010).

In agriculture sciences, nanotechnology helps to reduce the environmental pollution with the production of different types of pesticides carriers by using nano capsules and nano particles which have the capability of controlled or delayed delivery, more absorption, effective, ecofriendly, and the formation of nano-crystals which increases the capability of pesticides when applied in lesser dose. In near future nano particles will be widely used for the delivery of drugs or active ingredients to treat all the diseases associated with plants (caused by pathogens). There are a large number of nano materials which are used as pesticide or to deliver drugs. These nano materials include gold nano particles, polymeric nano particles and iron oxide nano particles which can be synthesized easily. Size, surface functionalization and shape of the nano particles may be modified and change the pharmacokinetic parameters of these nano particles.

Nano-scale based characterization of viral disease which include the multiplexed diagnostic kit development played very crucial role for the identification and detection of the specific stage of viral strain and utilization of drugs to prevent from the disease. The power and speed of disease detection increases by the utilization of nano-based diagnostic kits (Prasanna 2007). To make the "smart" agriculture system nano scale devices with unique properties could be used in near future. As an example, to identify the health issues of plant, nano devices could be used before these issues become notable to the farmer. These devices may have the capability to respond in different situations and to take proper action for the removal of disease. If these devices could not do so then they will alarm the farmer about the problem. In such a way the "smart" devices will serve as warning at very initial stage and offer prevention from the disease. These devices could be used in targeted and controlled manner to deliver the medicine in the same way when nano-medicine applied in humans for drug delivery. Developments in nano-scale medicine allow us the treatment of different diseases like cancer in animals because of the target delivery with high precision and it has become highly successful (Joseph and Morrison 2006). Therefore, they are also being considered to be targeted research area for plants.

5.4.1 Nanoscale Materials and Their Application

All the living organisms in nature, from the smallest to the largest depend on the nano-sized materials and the protein machines which are responsible for performing everything such as from the movement of flagella to the movement of a muscle. Nano-sized black carbon helps in the improvement of mechanical qualities of tires. The nano-sized silver nano particles start up the development of photographic film (Bhattacharyya et al. 2010). For the protection of environment, nanotechnology already has many great applications (Nowack 2009). Figure 5.2 describes the role of nanotechnology in agriculture.

5.4.2 Formulation of Nanopesticides

The formulation of any pesticide which includes the elements in the range of nm and claims the innovative properties are associated with these small ranges of size. Nano pesticides have a great variations in different products that's why we cannot consider all these in a single category. Nano pesticides may be consisting of:

1. Ingredients which are organic in nature and include polymers.
2. Ingredients which are inorganic in nature include metal oxides in different forms such as micelles and particles.

Like in the formulation of other pesticides the aims of nano formulations are:

1. To enhance the solubility of main ingredient.
2. To protect active ingredient against the early degradation.

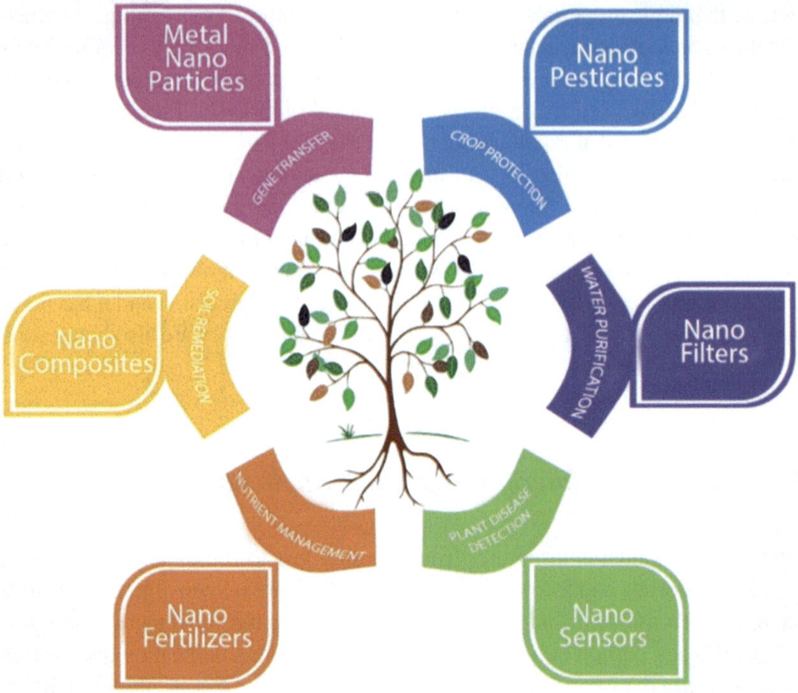

Fig. 5.2 Role of nanotechnology in agriculture

3. To release active ingredient in a slow manner.

At present time we cannot fairly assess the advantages and disadvantages of nano formulation due to the poor knowledge which results in avoiding the use of some nano pesticides.

5.4.3 Controlled Release of Pesticides Using Nanomaterials

Controlled release of pesticides with the help of nanomaterials is a new horizon to face challenge, rapid growth in the demand of food and energy as well as increased soil and air pollution. The controlled release of pesticide is a crucial step to achieve these goals of food, energy and environment sector (Sarlak et al. 2014).

Pesticide carriers are those substances which have the efficiency and stability to encapsulate or bind with active ingredients, as well as their controlled release properties such as an organic solvent and mineral clay which carries the active part of pesticides and transfer it to the target. The selection of material used as pesticide is very important and crucial for selection of carrier molecules (Yusoff et al. 2016).

To synthesize a pesticide, is a very difficult chemical process which need the experienced and trained chemist and a well-established, sophisticated laboratory. The primary process is to alter an organic molecule into pesticide. In order to formulate the pesticide products, ingredient, such as solvents, surfactants, emulsifiers, clay, propellants and water need to be added (Xu et al. 2016). But nanoparticles present easy solutions; a few commonly used carriers are as follows:

5.4.3.1 Nano-micellar Aggregates

Micelles are aggregates of electrically charged particles which occur in a colloidal solution. At the critical concentration these micelles are formed in solvent system. Micelles can be synthesized in different sizes ranging from 10 to 100 nm and their shapes depend upon their molecular weights. Nano-micelles are very small size micelles which have the particle size in the range of nano-scale, Nano-micelles are hydrophilic and hydrophobic (amphiphilic) colloidal structures formed by self-assembly of 5–200 nm monomers. For their ability to solubilize hydrophobic molecules in aqueous solution, nano-micelles have been an appealing carrier. Nano-micelles are commonly used for targeted delivery in agricultural, biological and pharmaceutical applications. The amphiphilic properties of the nano-micelles stimulate the insoluble molecules to be carried out and delivered to the specific target tissues (Rivera et al. 2016).

Nano-micellar aggregate is the most efficient way in which a micellar delivery system can maintain drug release. Typically, this strategy includes the formation of a drug conjugate with the hydrophobic portion of an amphiphilic polymer and the formation of micelles from this conjugate (Li et al. 2002).

For the release of drugs such formulations have two steps:

1. From the polymer the drug is removed by hydrolysis enzyme or other breakdown methods.
2. The drug is released out of the micelles through the diffusion of the drug, the former being typically the rate-limiting step.

A significant benefit of this technique is that because of conjugation, the drug stays in the micelle for a longer time (Zhang et al. 2016).

5.4.3.2 Nano Encapsulation for Controlled Release

The process of packing or enclosing the small particles of a substance inside the other material is called "Encapsulation". Encapsulated particles are intended and designed in such a way that, when these particles are subjected to a specific trigger they release their contents. Encapsulation can be helpful in making bioactive compounds and living organisms more stable such as bacterial spores being not interactive to the environment. Encapsulation can also be very helpful to ensure timely delivery of bio-pesticides, where they can be applied in the field. As an example, in a polymer a bacterial spores could be encapsulated which breaks down and open at the critical temperature, which ensuring that the spores activated at the same time when the pests arise (Bashir et al. 2016).

The amphiphilic copolymers (that can be synthesized from the polyethylene glycol and different aliphatic di-acids) are, self-assembled in aqueous media into nanomicellar aggregates, and are used to form controlled release (CR) formulations by using encapsulation techniques. The effectiveness of formulations can be increased by high solubilization power and low concentration of critical micelles (CMC). Various methods such as Dynamic Light Scattering (DLS), Infrared (IR) spectroscopy and Transmission Electron Microscope (TEM) can be used to characterize these formulations. The formulations which are formed for CR can be used in various plants and crops for effective pest management. When a commercial formulation is required for a practical implementation in the selected area, the use of materials is very crucial and should be compatible with the applications:

1. Ecofriendly.
2. Easily biodegradable.
3. Nontoxic waste products.
4. Cost effective.

The utilization of different biopolymers, i.e. polymers generated by natural sources, which simultaneously have excellent chemical and physical properties and still present in mild circumstances of biodegradation, is an interesting strategy to prevent the use of petrochemical derivatives that could be another cause of environmental pollution (Adak et al. 2012).

For the ideal drug action, the drug must be delivered as efficiently as possible to the required site of action. Targeting the drug to the site of action using either a prodrug or an advanced drug delivery scheme would be more beneficial. It not only enhances the therapeutic ability but also allows a decrease in the dose of the drug to be administered. As a result the unwanted toxic impacts are minimized. A delivery via colloidal drug delivery systems may be one possible way to deliver the drug to the targeted site. These devices give benefits in many fields like veterinary, medical, agricultural, and industrial applications, mainly because of their nano-size (Anamika et al. 2014).

5.4.3.3 Nanoparticles Used in Bio-pesticides Controlled Release Formulations (CRFs)

The most common forms of nanomaterials used for biocidal delivery in CRFs are:

1. **Nanospheres**: Aggregate with a homogeneous distribution of the active compound into the polymer matrix.

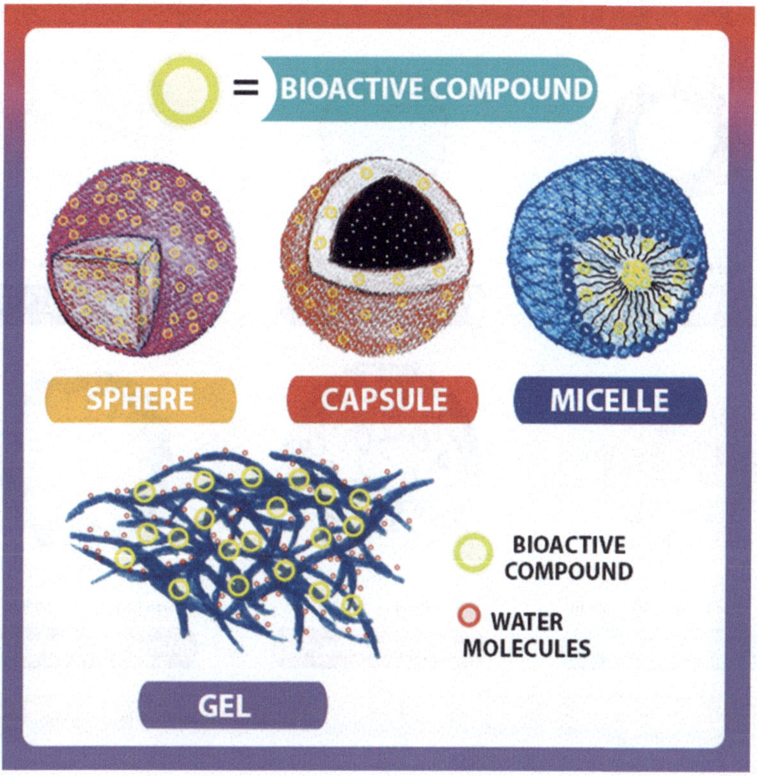

Fig. 5.3 Morphological description of Nanomaterials used in controlled release formulations

2. **Nanocapsules**: Aggregate in which the active compound is concentrated inside and surrounded by the polymer matrix near the center or nucleus.
3. **Nanogels**: Hydrophilic polymers (usually cross-linked) that can absorb excess amount of water.
4. **Micelles**: Aggregate synthesized by hydrophilic and hydrophobic molecules in aqueous solutions (Fig. 5.3).

5.4.4 *Organic Nanoparticles as Pesticide Carriers*

Polymerization properties of some organic compounds have led most of the scientists to create pesticide encapsulation from them as carrier transportation. Sarlak et al. (2014) used the multi-walled carbon nanotube (MWCNT) covered water-soluble dithiocarbamate fungicides as carrier with poly citric acid. The pesticide encapsulation was optimized with a stirring period of 30–80 min in the pH range of 6–8. In order to enhance the photo stability of various pesticides, polymeric nanoparticles were generated. The photo-degradation of the pesticide group of macrocyclic

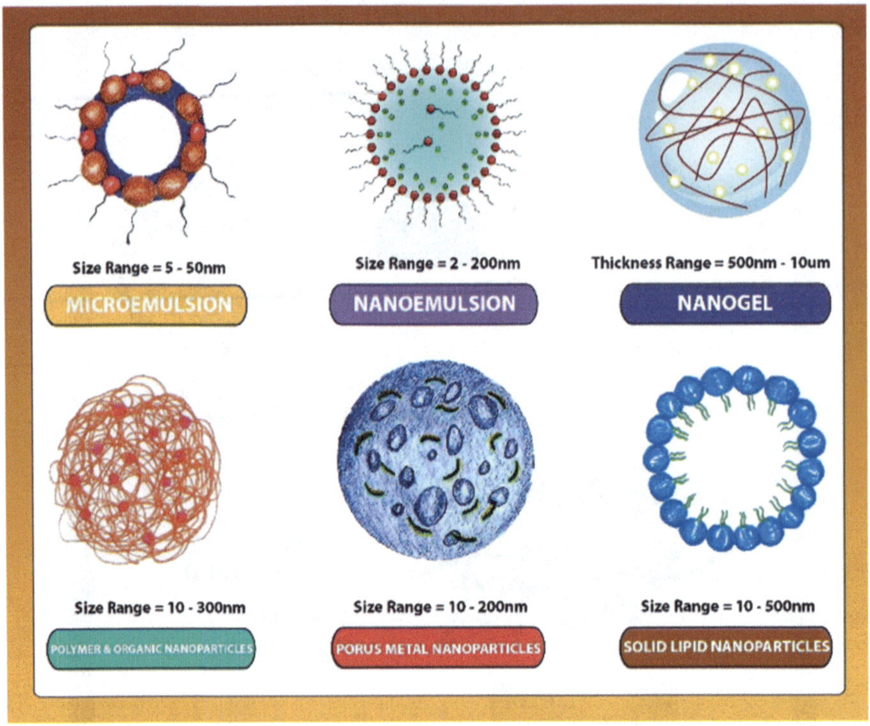

Fig. 5.4 Types of nano-pesticides carriers

lactones has been altered by encapsulation using polyacrylate polymer nanoparticles. The increase in the photo stability of the nanopesticide was apparently significant by the enhancement in insecticide activity against the *Helicoverpa armigera* which is cotton bollworm (Mishra et al. 2015). Photo stability of pesticides is advantageous as it increases the effect of applied dose so pesticides in lesser amounts are good in action. Polyethylene glycol (6000) nanoparticles were used to encapsulate plant-based pesticide oils in the ranges of 100–400 nm in size (Gonzalez et al. 2013). Figure 5.4 shows the types of nano-pesticides carriers.

5.5 Metal Nanoparticles as Pesticide

The nanoparticles having the particular biological and chemical properties can be used as a pesticide carrier or as a pesticide as the entire nano-sized metal entity. These metals can readily penetrate in the body of the insect and perforate the body. The antimicrobial and anti-parasitic properties of silver and silica nanoparticles are used in various agricultural and food processing applications for pest control. Metal nanoparticles and their insecticidal properties are given in Table 5.1. A comprehensive research was conducted on the strength and insecticidal activity of nanoparticles

Table 5.1 Applications of metal nanoparticles in pest management

Sr. No.	Metallic nanoparticles and preparations	Size (nm)	Target	References
1	Silver nanoparticles, extraction of chlorophylis	1–100	Mineralization of halocarbons	Manimegalai et al. (2011)
2	Gold nanoparticles, using the extracts of *Magnolia kobus* and *Diopyros kaki* leaf extracts	5–300	*Bipolaris sorokiniana* fungi	Al-Samarrai (2012)
3	Zinc oxide and Aluminum oxide nanoparticles, by the process of combustion	3–250	*Sitophilus oryzae* insects	Keratum et al. (2015)
4	Zinc nanoparticles by extraction	≤100	*Fusarium graminearum*	Dimkpa et al. (2013)
5	Cobalt nanoparticles, by plant extract	50–350	*Pseudomonas Aeruginosa* and *Escherichia coli*	Kuchekar et al. (2018)
6	Sulfur nanoparticles by Acid hydrolysis	10–100	*Escherichia coli*	Suleiman et al. (2013)

and terpenoidal silica nanoparticles by Rani et al. (2009). Nanoparticles surface layer expanded the storage capacity of pesticides to 180 days, and also ten times rise in insecticidal properties which were observed against two worms (tobacco cutworm and semi-looper worm). The silver nanoparticles were treated with Deltamethrin, a type II pyrethroid pesticide, and its mosquitocidal activity was tested against the *Aedes aegypti* dengue vector. The nanosilver-deltamethrin complex was formed in the size range of 20–300 nm and the deltamethrin insecticide activity was improved due to this complex formation. The use of metal nanoparticles leads to elimination of the accumulation of targeted pest. But, the excessive use of metal nanoparticles in plant cells can lead to the internalization of NPs and metal ions. This can lead to bioaccumulation and un-desirable bio-magnification of the metal in food chain (Jones and Hoek 2010). Metal nanoparticles which are non-specific in action can cause damage to the soil's useful microorganisms, and metal leaching can interrupt water ecosystem and aquatic invertebrate communities.

5.5.1 Nanopesticides and Environmental Concern

The pesticide's main ingredient is hazardous to environment because of several living and non-living variables. The biological efficacy and pesticide chemodynamics are influenced by abiotic conditions such as temperature, light, soil physicochemical aspects and salinity. Biotic factors such as microbes, crops and certain insects play a major part in the environmental degradation and persistence of pesticides. Microorganisms play a crucial role in maintaining soil chemical equilibrium and in soil and water decomposition of xenobiotic compounds. The selection of nanopesticide can resolve the issues posed with excessive use of the pesticide. But, there are numerous environmental dangers associated with the use of nano-products in pest control. Because of their large surface area, nanoparticles are reactive with different parts of the atmosphere, which can lead to an unexpected environmental outcome. Due to the variation in chemical ingredients and preparation process, the different types of nano-pesticides from emulsion to nano-dispersion have different environmental relationships. A clear knowledge of the safety of the environment, the future of nano-pesticide and its main ingredient is required before commercial use (Fenner et al. 2013).

5.6 Transformation of Nanoparticles in Environment

Nanoparticles form aggregates and clumps in the range of micron sizes because of the distinction in the ionic pressure between the environmental media and the surface of nano-particles. As described by Kah et al. (2014) Nano-pesticides based on metal active ingredients can undergo rapid environmental alterations by incorporating new chemical substances onto their surface. Combination of nanoparticles with different materials depends on their volume and nature, therefore, can increase the stability of toxic materials in environment as well. Polymeric nano suppliers and organic nanoparticles are capable of sedimentation with particles of soil and cause biotic degradation. The grouping of nano emulsion indicate modifications such as flocculation, phase separation, and emulsion creaming after too long storage, temperature and pH modifications (Sugumar et al. 2014). Figure 5.5 shows the transformation of nano-pesticides.

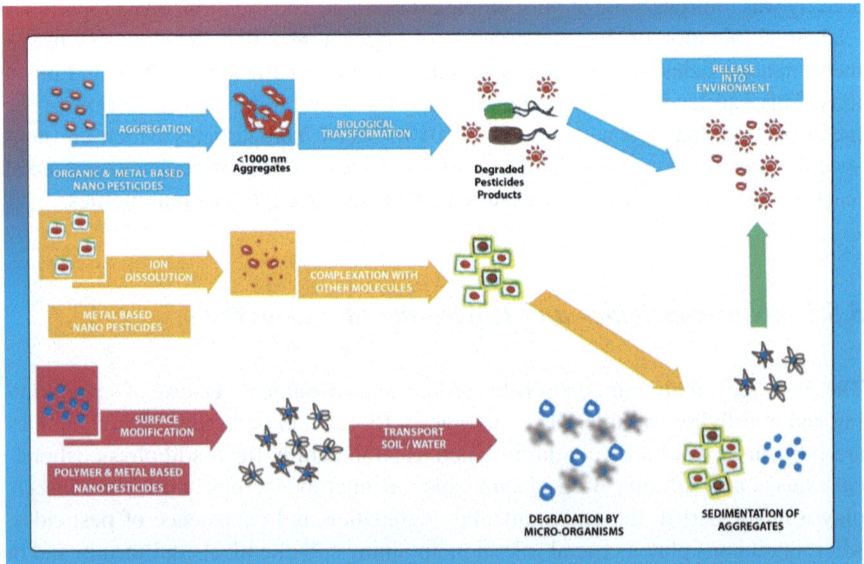

Fig. 5.5 Transformation of nano-pesticide in environment

5.7 Future Prospects

Applications of nanotechnology have great potential to alter production future in agriculture by enabling sustainable crop management. Nanotechnology offers much easier techniques of detecting threat of the pests and bio-remediating the environment. It can improve the productivity of agriculture by using nanoparticles as nanopesticides or by covering current pesticides with nanomaterials for smart delivery to plants. Both methods require extensive research work to define the optimized conditions for each crop with each nanoparticle and every pesticide. Nanosensors of particular type should be developed for each cash crop to deliver pesticide at particular timing and placement. Nanoparticles can also be used to degrade the remaining of pesticides from environment. Thus, nanotechnology will change farming methods in the future, including sophisticated pesticide management. Green revolution would be accelerated through nanoscience over the next 20 years. Nanotechnology could be used for the formation of new insect repellent and pesticides. Therefore, nanotechnology is regarded as one of the best possible solutions to issues in the food and agriculture sector.

5.8 Conclusion

In this century, increasing technologies will face a more suspicious and difficult challenges. These technologies have to make themselves clear not only for their benefits but also have to predict and describe potential hazards associated with them. Nanotechnology can be a major source of developing alternatives to minimize pesticide contamination. Nanotechnologies have demonstrated excellent ability to control the release pattern of the active ingredient of pesticides to make them more effective for long-term functionality that can overcome agricultural runoff and residual pesticide accumulation problems. Furthermore, nano-pesticide has shown enhanced solubility and stability of the active ingredient for efficient pest control. At present, nano-encapsulation is the most promising technology to protect host crops from insect pests. Thus, nanotechnology will improve agricultural output in the coming years, including pest management. But extensive research work is needed.

References

Adak T, Kumar J, Shakil NA, Walia S (2012) Development of controlled release formulations of imidacloprid employing novel nano-ranged amphiphilic polymers. J Environ Sci Health 47(3):217–225

Aktar WM, Sengupta D, Chowdhury A (2009) Impact of pesticides use in agriculture: their benefits and hazards. Toxicology 2(1):1–12

Al-Samarrai AHM (2012) Nanoparticles as alternatives to pesticides in management plant diseases: a review. Int J Sci Res Publ 2(4):1–4

Anamika D, Hussain F, Masram DT (2014) Synthesis, characterization, and antifungal studies of Cr(III) complex of norfloxacin and bipiridyl ligand. Bioinorg Chem Appl 2014:1–7

Bashir O, Claverie JP, Lemoyne P, Vincent C (2016) Controlled-release of *Bacillus thurigiensis* formulations encapsulated in light-resistant colloidosomal microcapsules for the management of lepidopteran pests of Brassica crops. Peer J 4:1–14

Bhattacharyya A, Bhaumik A, Rani PU, Mandal S, Timothy TE (2010) Nano-particles - a recent approach to insect pest control. Afr J Biotechnol 9(24):3489–3493

Brar SK, Verma M, Tyagia RD, Surampalli RY (2010) Engineered nanoparticles in wastewater and wastewater sludge – evidence and impacts. Waste Manag 30(3):504–520

Dimkpa C, Mclean JE, Britt DW, Anderson A (2013) Antifungal activity of ZnO nanoparticles and their interactive effect with a biocontrol bacterium on growth antagonism of the plant pathogen *Fusarium graminearum*. Biometals 26:913–924

Fenner K, Canonica S, Wackett LP, Elsner M (2013) Evaluating pesticide degradation in the environment: blind spots and emerging opportunities. Science 341(6147):752–758

Gonzalez H, Maestro LMP, Rosal J, Ramiro J, Caamano AJ, Carrasco E, Juarranz F, Sanz-Rodriguez F, Sole JG, Jaque D (2013) Nanoparticles for photothermal therapies. Nanoscale 5:7882–7889

Jones MC, Hoek EM (2010) A review of the antibacterial effects of silver nanomaterials and potential implications for human health and the environment. J Nanopart Res 12:1531–1551

Joseph T, Morrison M (2006) Nanotechnology in agriculture and food: nano-forum report, Institute of Nanotechnology, pp 1–13

Kah M, Machinski P, Koerner P, Tiede K, Grillo R, Fraceto LF (2014) Analyzing the fate of nanopesticides in soil and the applicability of regulatory protocols using a polymer-based nanoformulation of atrazine. Environ Sci Pollut Res 21:11699–11707

Keratum AY, Arab RBA, Ismail AA, Nasr GM (2015) Impact of nanoparticle zinc oxide and aluminum oxide against rice weevil *Sitophilus oryzae* (Coleoptera: Curculionidae) under laboratory conditions. ESPR 3(3):30–38

Kitherian S (2017) Nano and bio-nanoparticles for insect control: a review. Res J Nanosci Nanotechnol 7:1–9

Kuchekar SR, Patila MP, Gaikwad VB, Han SH (2018) Synthesis and characterization of silver nanoparticles using *Azadirachta indica* (Neem) leaf extract. IJESIRD 6(4):47–55

Li T, Ning F, Xie J, Chen D, Jiang M (2002) Preparation and morphologies of shell cross linked micelles based on commercial poly (styrene-block-ethylene-co-butene-block-styrene). Polym J 34(7):529–533

Manimegalai GS, Shanthakumar C, Rehaan M (2011) Pesticide mineralization in water using silver nanoparticles. Int J Chem Sci 9(3):1463–1471

Mishra M, Gupta KK, Kumar S (2015) Impact of the stem extract of *Thevetia neriifolia* on the feeding potential and histological architecture of the midgut epithelial tissue of early fourth instars of *Helicoverpa armigera* Hübner. Int J Insect Sci 7:53–60

Nowack B (2009) Is anything out there? What life cycle perspectives of nano-products can tell us about nanoparticles in the environment? Nano Today 4:11–12

Prasanna BM (2007) Nanotechnology in agriculture. ICAR National Fellow, Division of Genetics, I.A.R.I, New Delhi, India, pp 111–118

Ramsden JJ (2005) What is nanotechnology? Nanotechnol Percept 1:3–17

Rani A, Mun PV, Hande GLK, Valiyaveettil S (2009) Cytotoxicity and genotoxicity of silver nanoparticles in human cells. ACS Nano 3:279–290

Rivera FA, Villanueva DF, Concheiro A, Lorenzo CA (2016) α-Lipoic acid in soluplus polymeric nanomicelles for ocular treatment of diabetes-associated corneal diseases. J Pharm Sci 105(9):2855–2863

Sarlak N, Taherifar A, Salehi F (2014) Synthesis of nano-pesticides by encapsulating pesticide nanoparticles using functionalized carbon nanotubes and application of new nanocomposite for plant disease treatment. J Agric Food Chem 62(21):4833–4838

Sharon M, Choudhary AK, Kumar R (2010) Nanotechnology in agricultural diseases and food safety. J Phytol 2:83–92

Sugumar S, Clarke SK, Joyce N, Tyagi BK, Amitava M, Chandrasekaran N (2014) Nanoemulsion of eucalyptus oil and its larvicidal activity against *Culex quinquefasciatus*. Bull Entomol Res 104:1–10

Suleiman GM, Hussien NN, Marzoog TR, Awad HA (2013) Phenolic content, antioxidant, antimicrobial and cytotoxic activities of ethanolic extract of *Salix alba*. Am J Biochem Biotechnol 9(1):41–46

Xu H, Kim S, Sorek H, Lee Y, Jeong D, Kim J, Oh EJ, Yun EJ, Wemmer DE, Kim KH, Kim SR, Jin YS (2016) PHO13 deletion-induced transcriptional activation prevents sedoheptulose accumulation during xylose metabolism in engineered *Saccharomyces cerevisiae*. Metab Eng 34:88–96

Yusoff SNM, Kamari A, Aljafree NFA (2016) A review of materials used as carrier agents in pesticide formulations. Int J Environ Sci Technol 13(12):2977–2994

Zhang X, Lin W, Wen L, Yao N, Nie S, Zhang L (2016) Systematic design and application of unimolecular star-like block copolymer micelles: a coarse-grained simulation study. Phys Chem Chem Phys 18:26519–26529

Chapter 6
DNA Nanobiotechnology and Plant Breeding

Saadia Basheer, Khadija Rafiq, Muhammad Tariq Javed, Muhammad Shahid, and Muhammad Sohail Akram

6.1 Introduction

Nanotechnology refers to processes or applications which utilize materials (nanoparticles) with size range of 1–100 nm (in a single dimension). These nanoparticles have very unique properties in terms of their chemistry, physical appearance, reactivity and magnetic as well as optical effects. Generally, nanotechnology is characterized by the approach being utilized to achieve a task i.e. it may be 'top-down' or 'bottom-up'. The former is making nanoscale structures by utilizing the techniques of mechanizing, templating and lithographic, e.g. photonics applications in nanoelectronics and nanoengineering. Contrarily, 'bottom-up', (also called molecular nanotechnology), approach builds organic and inorganic materials into defined structures, atom by atom or molecule by molecule, often by self-assembly or self-organization, which can then be applicable in several biological processes.

Biotechnology is utilization of various advance technological tools and applications for better understanding and development of living things or biological molecules. Nanobiotechnology, therefore, is an interdisciplinary approach of utilizing the techniques of both nanotechnology and biotechnology for a larger benefit of mankind. Nanobiotechnologists have analysed and utilized various components of natural biological systems to design innovative nano-devices like biochips, molecular

S. Basheer
School of Biological Sciences, University of the Punjab, Lahore, Pakistan

K. Rafiq
Department of Botany, University of Gujrat, Gujrat, Pakistan

M. T. Javed · M. S. Akram (✉)
Department of Botany, Government College University, Faisalabad, Pakistan

M. Shahid
Department of Bioinformatics and Biotechnology, Government College University, Faisalabad, Pakistan

S. Javad (ed.), *Nanoagronomy*, https://doi.org/10.1007/978-3-030-41275-3_6

motors, nanocrystals and nanobiomaterials. Nano genomics-based methods have enabled breeders to achieve greater precision in breeding programs and have opened up new exciting opportunities for selecting and transferring genes. This has not only reduced the time needed to eliminate unnecessary genes, but has also allowed the breeder to access useful genes from distant plants.

The current chapter has highlighted the advancements in nanobiotechnology and its application towards plant breeding programs. Particularly, the role of nanotechnology, magnetofection, mesoporous silica nanoparticles, and clay nanoparticles in plant transformation/breeding has been discussed.

6.2 Mesoporous Silica Nanoparticles (MSNs)

Mesoporous silica nanoparticles are very small sized particles of silica (nano-size) with honey comb like channels and pores with great capacity of absorbing and carrying materials. The exceptional structural properties of organically functionalized mesoporous silica nanoparticles include thermally and chemically stable mesoporous structures, huge surface areas, highly organized surface properties, and adjustable pore sizes. These properties have increased their excellency for hosting various molecules of different sizes, shapes and functions. Mesoporous nanoparticles have appealed scientists for their application as carrier system specially their use in cancer treatment in animals.

MSNs are not toxic to plants and can be taken up by endocytosis in acidic lysosomes, therefore making MSNs unique candidates for delivery of drugs. Molecular gatekeepers, used for capping the MSNs, protect and help efficient internalization of MSNs into target. MSNs penetrate the roots by apoplastic and symplastic pathways through xylem (conducting tissue) towards the aerial parts of plants including stem and leaves (Sun et al. 2014).

In many systems which are based on non-porous nanoparticles for delivery in plant tissues and cells, particles of interest are restricted to nucleic acid, which usually adsorb on carriers exterior surface. For example, in gene gun system, which extensively uses DNA coated microparticles for bombardment of plant cells and tissues to transfer genes in these plants (containing cell wall). Such systems are used in many fields but co delivery of other molecules with nucleic acids is very difficult. For delivery of DNA and many other biogenic species, microinjection could be used. Although, this method is not appropriate for plant transformation due to its low efficiency and need to be replaced (Potrykus 1990).

In such situations, use of MSNs presents a solution. In MSN systems, silica nanoparticles are loaded in mesopores to stop the leaching of drugs. These drugs and imaging agents are then encapsulated with physical bound caps. Uncapping materials which enhance the breakdown of linkage between caps and MSN, are introduced in these molecules and they release pores. This design is feasible and is shown by using a disulphide antioxidant DTT (dithiothreitol) as a gateway for biogenic release of molecules in mammalian cells (Lai et al. 2003). Different research groups have synthesized a chain of MSNs with different functional groups which has helped in the investigations involving efficiency of MSNs. First investigation on MSN use was on protoplasts which are model systems in cellular and physiological studies (Sheen 2001). These mechanisms have been extensively studied and understood in animal cells where these studies were restricted to impermeable membrane dyes. Experiments were also performed for the mesophyll protoplasts of tobacco and they undergo endocytosis by use of Lucifer Yellow dye. Different types of MSNs are used for the uptake of nanoparticles.

Type-I MSN did not work for uptake of particles while Type-II MSN (Type-I MSN functionalized with Triethyl glycol/TEG) showed successful internalization and they were found in endocytic vesicles of cytoplasm during complete experimental duration. It is demonstrated that for plant endocytosis (cell) particle surface characteristics play a vital role. MSN concentrations per plant are observed to be low as compared with the animal cells. Size of endocytic vesicles between the range of 0.2 and 3 mm, which presents from 1 to 15 MSNs. Number of vesicles in each cell was highly varying from 1 to 20. This was the first example of plant cells (isolated) endocytosis for TEG coated nanoparticles. For cell biology and plant endocytosis studies, this was a new, unique and versatile MSN system because it caused no toxic effects in plants and proved to be totally safe and opened a new horizon of research.

For the evidence that MSNs could be used as DNA delivery agents in plant cell, a plasmid containing GFP (green fluorescent protein) gene under constitutive promoter control was used. Optimal coating ratio for DNA and MSNs was 1/10. For Type-II MSNs, DNA complex formed successfully and this proved that stable DNA and type-II MSN complexes were formed because no free DNA was detected in solution after incubation. Restriction enzymes did not digest the DNAs bound to Type-II MSN. GFP expression (transient) could be observed after protoplast incubation with MSN coated with DNA. Type-II MSNs could be identified in whole GFP expressing cells. Standard PEG (Polyethylene glycol) mediated transformation (protoplast) could be achieved with 40–90% transformation efficiency when 1–2 mg of DNA/106 cells is used (Torney et al. 2007).

It is found that Type-II MSN coated by DNA serves as an excellent delivery system of DNA to protoplast and could make DNA, approachable to transcription machinery which leads to transgene expression. For genetic engineering of plants desired targets are intact plant tissue. It was tested that MSNs could be introduced into plant cells with gene gun method. Attempts for MSNs bombardment did not result in successful transformation and there were no MSN in the cells, may be because of lower density of silica nanoparticles.

Another type, Type-III MSN was introduced to overcome this issue, in which mesopores were encapsulated with surface-functionalized, gold particles, which served not only as biocompatible covering agent (Shukla et al. 2005), but also increased the weight in every MSN to enhance the density of the resulting material. On tobacco cotyledons, GFP expressing foci can be seen which are coated with Type-III MSN. By the use of gene gun method, gold nanoparticle-capped with Type-III MSN could be used for delivery of DNA in intact plant cells and tissues.

Mesoporous structures have more efficiency of loading larger amounts of biogenic moieties. These include chemicals which are membrane impermeable or incompatible to growth media. In comparison, the gold particle DNA delivery system has DNA coated with solid gold materials. Maize immature embryos are also bombarded by Type-III MSNs. GFP expression was also observed in callus portions from a culture grown in non-selective medium few days after bombardment. Gold nanoparticles may be aggregated on MSN surface. It is concluded that the DNA can form stable complexes on Type-III MSN surface; this gene system could be applicable for both stable transgenic and transient plant materials.

Most beneficial feature of MSN technology is the potential to be delivered in varying biogenic species at the same time to target site and discharge the encapsulated chemicals in control manner. For testing whether controlled discharge for the animal cell (Radu et al. 2004) are also applicable for plants, a transgenic tobacco was generated which contained inducible promoter control GFP gene. GFP expression in plants was observed only in presence of a chemical inducer, beta-oestradiol (Zuo et al. 2000). Average fluorescent foci are counted for different transgenic events which are germinated on medium containing DTT. It is concluded that gene present inside the plant genome can be activated through chemicals delivered in cells, which directly leads to controlled discharge in planta by the use of this MSN system.

Finally for demonstration of MSN system transparency and delivery of both chemicals and gene simultaneously, experiments were performed where plants were bombarded with MSNs loaded with beta-oestradiol along with gold nanoparticle coated with GFP marker gene. The results described that DNA molecule delivery of a marker gene along with the chemical at same time is possible. The technique releases the encapsulated chemical by controlled manner for trigging expression of co-delivery of transgene in cells. Currently nanotechnology is basically applied to animal and medical sciences but it has been demonstrated that it could also be used in plant sciences for crop improvement and genomic studies (Bharti et al. 2015; Torney et al. 2007). During material synthesis direct functionalization of silica surface is allowed by the co-condensation process, which is preferential for internal mesopores surface. Porous silica materials are the most beneficial compounds which provide pathways for the challenging diseases treatments. Many broad range advantages of silica include its versatility, biocompatibility, pore volume, biodegradability, porous space for guest molecules of different pore sizes, and surface charge control, and their dispersion throughout plant body.

Another application of MNPs in plants is to produce magnetoferon pollens (pollen with external DNA transferred by magnetic nanoparticles). Pollen of most crops have surface apertures with diameter of 5–10 μm with a pollen wall reduction or absence. It is confirmed that cotton pollen has aperture structures where pollen wall is thin and has high permeability which helps in delivery of exogenous genes into the pollen. Pollen magnetofection is tremendously mild and cause no damage to pollen. Pollen tube growth and germination of magnetofected pollens showed high pollen viability. Pollen germination showed higher viable pollen. It is because of the sizes of MNPs that are smaller than 200 nm and could pass by the pollen apertures.

This whole procedure involves a mixing of MNP–DNA complexes and pollen, effectively reducing pollen perturbation.

MNPs exhibit advanced properties in protection and concentration of exogenous nucleic acids. DNA molecule adsorption on the surface of MNPs is demonstrated clearly in mechanism of assembly of MNP/DNA complexes through morphology characterizations. MNPs can efficiently transfer exogenous gene in mammalian somatic cells and have high stability in gene expressions. Different structures of MNP/DNA complexes are produced with varying ratios of MNP/DNA. Due to these characteristics, their biological performance is affected, leading to changes in dose–response relationships and cytotoxicity. Application of MNPs in efficient magnetofection and their gene delivery systems and functions are elaborated effectively (Zhao et al. 2014).

6.3 Clay Nanosheets

Plant pests and pathogens are affecting the global food security by reducing the crop yields about 30–40% per year (Borges and Martienssen 2015). Increased global warming is another environmental problem responsible for moving the causative agent of diseases into new territories (Bebber et al. 2013). Disease control programmes depend on plant genetic resistance and/or transgenes in combination with insecticidal and fungicidal sprays. There is a need of versatile strategies for managing the crop health which in turn is driven by greater crop production, pesticide resistance and toxicity problems among different climate ranges. RNAi (RNA interference) is a conserved eukaryotic mechanism present in both plant and animal kingdoms (Baulcombe 2004). It plays an important role in growth, development, and defence mechanisms of host against viruses and transposons.

RNAi, with the help of its DCL (DICER LIKE) enzyme activity, processes dsRNA into small interfering (si) RNAs. These siRNAs are incorporated into RNA-induced silencing complex which ensures that they particularly degrade RNA shared sequences resembling with the induced dsRNA. RNAi pathway confers resistance to transgenic diseases against pathogenic microorganisms in plants (Duan et al. 2012; Gordon and Waterhouse 2007). Topical application of dsRNA emerged as an attractive alternative for pest control in comparison to conventional transgenic methods. It is concluded that exogenous application of dsRNA could develop a new environmentally protected biotechnological method that aims to protect plants against viral diseases. However, there are some limitations with dsRNA topical application that could be addressed through advance nanotechnology techniques (Robinson et al. 2014; Tenllado et al. 2004). Encouraging research is being conducted for the application of dsRNA topically for plant virus control (Lau et al. 2014). The short period of post spray protection of dsRNA from virus is the limitation in its practical application (Gan et al. 2010). There is a preliminary information about BioDirect technology, as an RNAi spray application, from Monsanto, but no details emerged on surface.

There are enormous questions regarding dsRNA fragment delivery into the plant cells, mechanism of uptake in dsRNA into plants, and the issues regarding stability of the dsRNA applied topically to withstand all environmental conditions and facilitate long term protection against viruses. For this purpose potential applications of clay nanoparticles have been investigated. These sheet-like clay nanoparticles are composed of positively charged LDH (Layered Doubled hydroxide) nanosheets that contain dsRNA for prolonged and efficient protection against viruses. LDH materials are found in nature either because of precipitation in salt water bodies or due to weathering of basalts. LDH nanosheets belong to the family of inorganic layered materials (Xu et al. 2006) containing mostly divalent and trivalent metal ions.

A striking alternative to RNAi (transgenic RNA interference) is the use of pathogen-specific dsRNA (double-stranded RNA) for resistance against virus in plants. For practical application of this technology, a major limitation is the instable nature of naked dsRNA. It is demonstrated that dsRNA could be loaded on non-toxic, designer and degradable LDH (layered double hydroxide) clay nanosheets. Once dsRNA has been loaded on LDH, these do not wash off, sustained discharge and could be found on sprayed leaves even after many days of application. Evidences are being provided for degradation of LDH, silencing of homologous RNA and dsRNA intake in plants by topical application. Radically, single spray of this RNA loaded on BioClay (LDH) protected the plants from virus attack at least 20 days when assessed on sprayed and newly born unsprayed leaves. This invention translates nanotechnology made for the delivery of RNA interference for its use in environment friendly easily adaptable topical spray of human therapeutics.

Efficient laboratory protocol has been developed for the production of homogenous stable positive charge bearing LDH nanosheets which have particle size from 80 to 300 nm. Large dsRNA could be loaded on LDH nanosheets and form complexes known as BioClay (dsRNA–LDH). The BioClay dsRNA could be detected even after 30 days of topical application on leaf surface and is protected from nuclease activity. Under ambient environment, BioClay provides sustained discharge of dsRNA on leaf surface. Over a time period, LDH nanosheets could be degraded, most importantly BioClay sprayed topically provides RNAi based protection to unsprayed newly emerging and sprayed leaves for a longer duration of time after one spray. There are evidences for the uptake of dsRNA for triggering RNAi against homologous RNA.

6.3.1 Degradation of LDH and Release of dsRNA

Moisture and CO_2 can gradually breakdown LDH in a biochemical residue and release the loaded biomolecule. This test was performed on leaves of *Nicotiana tobbacum* by placement of LDH suspension droplets and incubated at 27 °C with 95% humidity in natural environment or with 5% CO_2. After a week of exposure LDH residues on the leaf showed decrease in Al (28%) and Mg (22%) amounts. dsRNA loaded into the leaves was also tested in the solution by incubation of CMV2b-dsRNA–LDH complex for 7 days either in normal environmental conditions or 5% CO_2. The residues were tested after incubation through Northern blot analysis. After comparison between normal atmospheric and 5% CO_2, the former atmospheric conditions showed a decrease in LDH-bound CMV2b-dsRNA. This indicated the LDH dissolution, at pH 3.0, together with release of dsRNA.

These conclusions indicated that degradation of LDH nanosheets occurred through the formation of carbonic acid when CO_2 in the atmosphere reacted with water film on leaf surface, facilitating the slow release of dsRNA.

6.3.2 *Uptake of dsRNA into Plant Cells and Induction of RNAi*

The ingestion of dsRNA or small regulatory RNAs into the plant cells is a crucial step in topical application of dsRNA. Different Cy3 fluorophore alone as well as LDH, Cy3, CMV2b and dsRNA combinations were applied to the leaves of *Arabidopsis thaliana* and then observed under confocal microscopy after 2 days of application for Cy3 uptake in leaf. There was no uptake of Cy3 alone and in combination with LDH and LDH was found in xylem with CMV2b-dsRNA–Cy3. CMV2b-dsRNA–Cy3-LDH demonstrated that Cy3 is found in mesophyll abundantly. Many observations showed that dsRNA could be taken up by plant cells by passive diffusion or by active transport. GUS (β-glucuronidase) reporter system could be induced for further confirmation for RNA uptake and RNAi induction in Arabidopsis.

There was no specific difference in GUS expression in seedlings which were treated with empty vector, LDH (control), water, non-specific CMV2b-dsRNA. Difference was found in seedlings treated with GUS-dsRNA or with GUS-dsRNA–LDH in comparison with control. Naked dsRNA and dsRNA with LDH both were taken up by seedlings and they regulated the post transcriptional down regulation of expression of GUS.

Results of GUS system showed that dsRNA enters when applied to whole plant body and transgene expression was silenced by RNAi pathway (Mitter et al. 2017). Recently there are two reports regarding uptake of dsRNA through roots (Jiang et al. 2014; Li et al. 2015) and one through flower bud (Lau et al. 2015). dsRNA moves systemically inside the whole plant has been confirmed by confocal microscopy.

6.3.3 *Stability of dsRNA Loaded into LDH*

For long lasting RNAi-mediated protection from viruses, the applied dsRNA (loaded into LDH) must not reside only on the leaf surface for prolonged duration than naked dsRNA, but they should be stable enough to sustain harsh field conditions like rain, nucleases and sunlight. The LDH nanosheets capability of protecting dsRNA from degradation could be observed by testing with confocal microscopy and Northern blot analysis. LDH nanosheets could protect dsRNA from nucleases, resist washing and mediate sustained discharge on leaf for extended duration. The fluorescence was not observed after rinsing in case of Cy3 alone, while Cy3-LDH exhibited fluorescence both before and after rinsing. Most of the CMV2b-dsRNA–Cy3 washed away and CMV2b-dsRNA–Cy3-LDH remained on the leaves after rinsing.

6.4 BioClay Spray for Crop Protection

The effectiveness of BioClay topical application, compared with naked dsRNA, to mediate RNAi protection mechanism against plant pests was tested by local lesions as well as systematic infection assays. It was confirmed that CMV2b-dsRNA spraying could induce virus resistance (sequence specific). BioClay application resulted in low necrotic lesions on the plants sprayed with CMV2b-dsRNA or in combination with BioClay on both time intervals. Mechanical inoculation process in cowpea (*Vigna unguiculata*) is a limitation that cowpea cannot be exposed to spray treatments after 5 days because they themselves degrade the old leaves without any treatment. There are many demonstrations regarding viral protection from topical spray of dsRNA application (Tenllado and Dıaz-Ruız 2001). Results from earlier experiments clearly showed that BioClay gives longer protection to plant cells against a large set of viruses as compared to the naked dsRNA. Collectively, it is concluded that CMV2b-BioClay application results in reducing the viral infections in long and short both terms, while CMV2b-dsRNA is only efficient in short term protection. CMV2b-specific RNAs could also be detected in newly emerged unsprayed leaves in whole plants which were sprayed with CMV2b-BioClay (Mitter et al. 2017).

It has been demonstrated that BioClay are viable, and is a new LDH nanocarrier system for the dsRNA for delivery of RNAi and it is a stable crop protection mechanism (spray application) which negates conventional transgenic procedures. Up till now, the field for application of nanoparticles for the delivery of RNAi is human therapeutics predominantly (Wittrup et al. 2015; Kanasty et al. 2013) with minimum research in plant applications because of cell wall barrier (Jiang et al. 2014). Previous studies indicated that LDH (loaded with siRNA) showed gene silencing expression in cells and effective delivery of siRNA in mammalian cells (Ladewig et al. 2009). BioClay platform efficiency can be attributed by the elegant design and functions of these LDH nanosheets for loading dsRNA, as these can strongly stick to the leaf. These are not affected even after vigorous dipping, and under atmospheric conditions they increase the stability of dsRNA for longer duration. For the release of dsRNA, carbonic acid is produced on leaf, which in turn degrades LDH nanosheets.

6.5 Diffusion-Based Biomolecule Delivery

Plant biotechnology helps in provision of resources for world's leading challenges which mainly include food security, higher demands for energy and pharmaceutical manufacturing. Genetic enhancement of plants is employed to produce high resistance against diseases, insects and herbicides. Bioengineered plants provide excellent biofuels and help generate small drug molecules and proteins (recombinant). Instead of several advancements in biotechnology, genetic transformation of most of the plant species is difficult to achieve. Delivery of biomolecules in plant cells is limited and a big problem towards effective genetic transformation. Plant cell walls are rigid, and are multilayered so acts as a barrier for plant genetic transformation. Biolistic particle and agrobacterium delivery techniques are the most commonly used techniques for gene delivery into plants, but these techniques have many limitations i.e. they are narrow in range in plant species and are tissue specific (Baltes et al. 2017).

Nanomaterial based delivery systems are understudied in plant cells; however, their potential use and delivery system is well elaborated in animal cells (Demirer and Landry 2017). Researchers have described the nanomaterials uptake in plant cells and most of it is non-essential cargoes, delivery into plant cells is carried out by mechanical aids, and use of protoplast cell cultures for the entry of nanoparticles instead of thick walled plant cells (Silva et al. 2010). For the delivery of functional cargoes into the plant cells, avoiding mechanical aid, nanomaterial internalization is demonstrated by LDH (layered double hydroxide), silicon carbidewhiskers, MSNs (mesoporous silica nanoparticles), clay nanosheets and DNA nanostructures and origami (Demirer et al. 2019).

Currently, there must be a plant transformation system which can facilitate highly efficient DNA delivery, avoiding unnecessary transgene integration, and work in a species independent way. Long-standing issue of DNA delivery into the plant cells has been studied in model and non model plants by nanomaterials. High aspect nanomaterials like CNTs (carbon nanotubes) have been studied to traverse extracted chloroplast passively, and plant membranes on several particular merit scales: exceptional tensile strength, high aspect ratio, high surface area to volume ratio with strong biocompatibility. Biomolecules, when bound to CNTs, are protected from degradation and cellular metabolism (Wu et al. 2008), giving higher biostability as compared with free molecules.

Moreover, SWCNTs (single-walled carbon nanotubes) have strong NIR (near infrared) fluorescence, inside the tissue transparency window thus advantage from reducing photon scattering, allowing nanoparticles delivery in plant tissues. Previous integration of CNTs is limited only to exploratory studies on CNT compatibility and small molecule sensing in plant tissues by the introduction of CNTs complexed to synthetic (fluorescent) polymers or dyes (Wong et al. 2016). It is a CNT-based platform which provides further advances in the field of nanoparticle based plant transformation. This validate and accurate platform delivers DNA in model and crop plants with greater efficiency, no toxicity and without transgene integration and mechanical aid. This combination of highly efficient features is not possible with conventional transformation approaches. These CNTs have delivered DNA into mature *Triticum aestivum*, *Gossypium hirsutum*, *Nicotiana benthamiana* and *Eruca sativa* leaves and generated protein expression. Studies have established that low size exclusion limit of CNTs are a promising solution for the problems regarding biomolecules delivery in a non-integrating and specie independent way that could be a new source of plant biotechnology applications (Demirer et al. 2019).

6.6 Grafting DNA on CNT (Carbon Nanotubes) Scaffolds

For loading of GFP encoding plasmids or linear PCR amplicon on MWCNTs (multi-walled carbon nanotubes) and SWCNTs (single walled carbon nanotubes), two distinct grafting methodologies were performed in an earlier research. One method is DNA grafting for direct loading of DNA on surface of CNTs through dialysis. Initial coating of CNTs is performed with sodium dodecyl sulfate (SDS) surfactant. During dialysis, through a dynamic ligand exchange method, DNA is adsorbed on the surface of CNTs and the SDS desorbs from its surface and leaves the dialysis membrane. By π–π stacking interactions, dsDNA vectors are grafted on the CNTs. By SWCNT near infrared fluorescence emission spectra adsorption of DNA on CNTs surface is confirmed by solvatochromic shift, it exhibits a DNA adsorption induced change in CNT dielectric conditions. These dialysis aliquots of CNTs (coated with SDS) without DNA exhibit rapid precipitation of CNT and have no NIR fluorescence, and confirms the replacement and desorption of SDS by DNA in dialysis aliquots with DNA.

Another technique used for the delivery of CNTs is by electrostatic grafting known as DNA grafting, where carboxylated CNTs are modified covalently to a cationic polymer for carrying positive charge. Negatively charged DNA vectors (DNA-CNT) are then incubated with positive charge carrying CNTs. After each step, height of CNTs increases, indicating that PEI attachment and DNA adsorption is performed which can be verified by AFM. This attachment is also revealed through zeta potential measurement. DNA conjugates 'DNA-CNT' which are prepared through electrostatic grafting have greater loading efficiencies of DNA then the conjugates of dialysis method. When loaded with similar amount of DNA, the loading efficiency of DNA-CNT conjugates of electrostatic grafting were 100% but the loading efficiencies of dialysis method were 50–70% (Demirer et al. 2019).

Intracellular stability of DNA-loaded PEI–CNT is measured through incubation with proteins as plant intracellular conditions are at complete protein concentration. After this incubation half of DNA remains on the surface of nanoparticles this indicates the stability of DNA into the plant tissues. By comparing free DNA and DNA with PEI–CNTs, this conjugated DNA is protected from endonuclease degradation when proteins extracted from leaves is incubated with it. DNA-PEI CNTs remain (50–70%) stable and intact after some days after incubation but 100% of free DNA is degraded. This could be validated through smTIRF (single-molecule total internal reflection fluorescence) microscopy (Demirer et al. 2019).

6.6.1 DNA Delivery into Mature Plants with CNTs

Gene expression in laboratory species are tested with arugula, cotton for their transformation in addition to model plant *N. benthamiana* on this platform for the demonstration of applicability of the procedure. Moreover, protein expression and gene delivery experiments are carried out in wheat plants for signifying the applicability of this protocol in monocot species besides dicot plants. Cy3-tagged DNA-CNTs were delivered to leaves and the infiltrated leaf tissue were observed with confocal microscopy for the assessment of nanoparticle fate.

DNA–CNTs internalization into leaf cells (mature) is confirmed by the TEM (transmission electron microscopy) and NIR imaging of these CNTs within leaf tissue by taking benefit of intrinsic NIR fluorescence of SWCNTs. DNA–CNTs infiltrated shows that CNTs amount reduced per leaf area, nearly 50% after some days, presumbly due to leaf expansion and cell division thereby resulting in CNT dilution. It is also assessed that DNA–PEI–CNTs can also internalize into chloroplasts. DNA–PEI–CNTs which have positive Zeta potential have internalized into the chloroplasts (extracted chloroplasts). DNA–PEI–CNT chloroplasts internalize conclusions are in accordance with lipid exchange envelope penetration model, which shows the nanoparticle internalized into small dimensions below ~20 nm and positive zeta potential value above ~+30 mV or below ~−30 mV (Wong et al. 2016; Giraldo et al. 2014). GFP fluorescence arises from the full leaf thickness was revealed by the Z-stack analysis of the DNA–CNT treated leaves' fluorescence profile. This confirms that full leaf profile is penetrated and diffused by CNT nanocarriers. In the absence of nanocarriers, there is no GFP expression in leaves with free DNA vectors. Under different formations of nanomaterials, the effectiveness of GFP expression and nanocarrier internalization differs substantially. Arugula leaves quantitative fluorescence intensity analysis through confocal images indicates that electrostatic grafting results in more significant GFP expression of DNA–CNTs as compared to GFP expression bring on by DNA–CNTs formed by π–π grafting through dialysis. It is demonstrated that gene expression mediated by CNT in mature leaves is transient and independent of target plant species. It was recorded that GFP expression was transient and genes which diffused into plant cells through nanocarriers did not integrate into nuclear genome of the plant (Demirer et al. 2019).

6.6.2 DNA Delivery into Isolated Protoplasts with CNTs

Nanocarriers have the ability to deliver plasmid DNA and to trigger gene expression into unique plant system of the isolated protoplast. Plant mesophyll protoplasts could be internalized through CNT formulations (Lew et al. 2018). Protoplasts are extensively used for the enhancement of output of plant genetic screens and in the production of recombinant proteins, consequently taking advantage of a facile, highly efficient and specie independent platform (Schaumberg et al. 2016).

Internalization of nanomaterial is confirmed when imaging of protoplasts is observed after incubation of protoplasts with Cy3–DNA–CNTs. It was confirmed that nanomaterial penetrated the cytoplasm and nucleus of the protoplast.

6.7 Magnetofection

Plant transformation techniques to produce genetically modified crops with versatile characteristics are of major concern at present. Conventional breeding methodologies have been extensively used for the production of genetically modified plants through laborious and time consuming regeneration cultures. New and advance technologies are established which are of vital importance for improvement in transformation of DNA in GM plants. One of such technology, pollen magnetofection, involves the production of transgenic seeds without traditional regeneration processes. This method is based on the pollen mediated transfer of selected genes (Schreiber and Dresselhaus 2003; Stöger et al. 1992). Active DNA is transformed to pollen during pollination and fertilization. Direct generation of transgenic seeds could be achieved through pollination by exogenous DNA transformed pollen (Touraev et al. 1997; Ohta 1986). DNA and RNA could be exploited for the production of desired protein from cells, endogenous gene expressions and even for repairing the defective genes. Their insertion in the target cells is highly challenging. During previous 10 years, nano-magnetic techniques for the delivery and targeting of nucleic acids have been established, which are known as magnetofection (Plank et al. 2011). In pollen-based magnetofection, Fe_3O_4 MNPs, coated with positively charged polyethyleneimine, were employed as DNA carriers for holding the negatively charged DNA, forming the MNP–DNA complexes. This MNP–DNA is then directed to pollen in the presence of an applied magnetic field.

Magnetofection methodologies are based on principles of Widder and others in late 1970s for delivery of magnetically targeted drugs. Transfection through magnetic microparticles was first established *in vivo* in mice and *in vitro* in C12S cells by Cathryn Mah, Barry Byrne and many others in 2000. Basically this technique is established by coupling magnetic nanoparticles and genetic material. *In vitro* magnetofection is actually the introduction of particle/DNA complex in cell cultures where earth magnets have established field gradient which increases the speed and sedimentation of the transfection (Dobson 2006). Magnetofection is taken as an ideal and efficient approach for gene transfer which uses magnetic force for insertion of DNA associated with magnetic nanoparticles (MNPs) in target cells (Scherer et al. 2002). Magnetofection have been applied currently in animal, human medical research and in plants for various purposes (Wang et al. 2014). Cell wall in the plant cells are resistant to exogenous DNA insertion unlike bacteria and mammalian cells (Torney et al. 2007).

Paramagnetic particles are exploited as drug carriers in magnetic targeting; it directs its accretion in target tissues with full magnetic field strength and is successfully used in cancer treatment (Lübbe et al. 1996). *In vitro* and *in vivo* efficacy of gene vectors could be improved by the application of this principle in transfection (Scherer et al. 2002). Both these goals need rapid transfection kinetics, possible vector targeting to chosen area and improvement in characters of dose–response. Magnetofection is an advanced tool for the solution of problems regarding slow vector accumulation which results in low concentration of vector on the target tissues and reduces gene delivery (Luo and Saltzman 2000). Physical association of gene vectors to super-paramagnetic nanoparticles (coated with polycation) enhances vector accumulation/concentration and thus, gene delivery. Magnetofection strongly benefits from this fact that individual system modules can be independently optimized and variants are assembled in combination, and thus important towards specific application.

6.7.1 Genetic Transformation Mediated by Pollen Magnetofection

Introduction of the desired gene into the pollen grain is necessary for the subsequent use in pollination and fertilization. Zygote is produced by fusion of egg and sperm cells at fertilization. So it is integral to integrate transgenes into the specific generative cell of the pollen. Because of thick cell walls of pollen grains it is very difficult to introduce genes into mature pollen cells with conventional breeding transformation methods. Initially experiments failed to produce viable transgenic plants after pollen treatment with exogenous DNA after pollination and fertilization (Booy et al. 1989; Sanford et al. 1985). Pollen transformation is carried out on the basis of assumptions that DNA uptake of pollen lead to integration of transgene in germline with consequent transmission into progeny (Eapen 2011).

In breeding of transgenic plants by pollen magnetofection, cotton was used as a representative species due to its high economic value but difficult regeneration. pBI35SBTΔ αCPTI plasmid containing magnetofected pollen was collected and then through artificial pollination BTΔα-CPTI-kanares (insect-resistant transgenic seeds) were produced. Different generations of cotton transgenic plants were screened from seedlings of genetically modified seeds. Very interesting feature of genetic transformation of target cells is that transgene incorporation by genome doubling creates exclusive opportunity for the immediate generation of plants homozygous for the transferred gene (Kumlehn et al. 2006). Nanoparticle and nucleic acid complex formation is necessary to complete gene delivery. Complex formation is fundamental for exogenous gene application, expression that depended on DNA and RNA concentration assembling mechanism and ratio. Two main mechanisms of assembly, adsorption and encapsulation, depend on modified functional groups and size of vector (Zhao et al. 2014).

Pollen magnetofection is a platform for high profile technology in crop genetic transformation and provides efficient in-field operation. It can evade the tiresome processes of tissue culture and develop transgenic plants from transformed seeds in short intervals of time. Through pollen magnetofection production of scaled-up germplasm transformations are possible and it creates new breeding techniques for genetically modified plants. This technology will lead to the development of crops particularly those which are difficult to transform using current methods. This will help farmers and public to achieve better crops in suitable yields.

Formation of multifunctional magnetic nanoparticles is one of the most attractive research areas in recent era. MFMNPs have magnetic properties and functions and it has been demonstrated to have great role as multimodality imaging probes. Multifunctional surfaces allow them to conjugate with other biological and drug molecules, for making target specific diagnostics and therapeutics (Hao et al. 2010).

6.8 Conclusion

Plant breeding programs must be speed up in order to get varieties with more production, nutritive value and higher resistance to pests. In a slow tempo of traditional techniques, advanced formulations of nanobiotechnology like magnetoferon pollen, mesoporous silica nanoparticles, carbon nanotubes and clay nanosheets can play role by enhancing the pace of DNA transfer to various host systems, importantly the plants. These methods give more precision and accuracy with greater stability. Extensive research work is needed in this field as the system is still in its infancy but it could be the best choice for generation of transgenic organisms.

References

Baltes NJ, Gil-Humanes J, Voytas DF (2017) Genome engineering and agriculture: opportunities and challenges. In: Progress in molecular biology and translational science, vol 149. Elsevier, Amsterdam, pp 1–26

Baulcombe D (2004) RNA silencing in plants. Nature 431:356–363

Bebber DP, Ramotowski MA, Gurr SJ (2013) Crop pests and pathogens move polewards in a warming world. Nat Clim Chang 3:985–988

Bharti C, Nagaich U, Pal AK, Gulati N (2015) Mesoporous silica nanoparticles in target drug delivery system: a review. Int J Pharm Investig 5:124–133

Booy G, Krens F, Huizing H (1989) Attempted pollen-mediated transformation of maize. J Plant Physiol 135:319–324

Borges F, Martienssen RA (2015) The expanding world of small RNAs in plants. Nat Rev Mol Cell Biol 16:727–741

Demirer GS, Landry MP (2017) Delivering genes to plants. Chem Eng Prog 113(4):40–45

Demirer GS, Zhang H, Matos JL, Goh NS, Cunningham FJ, Sung Y, Chang R, Aditham AJ, Chio L, Cho MJ (2019) High aspect ratio nanomaterials enable delivery of functional genetic material without DNA integration in mature plants. Nat Nanotechnol 14:456–464

Dobson J (2006) Gene therapy progress and prospects: magnetic nanoparticle-based gene delivery. Gene Ther 13:283–287

Duan C-G, Wang C-H, Guo H-S (2012) Application of RNA silencing to plant disease resistance. Silence 3:5

Eapen S (2011) Pollen grains as a target for introduction of foreign genes into plants: an assessment. Physiol Mol Biol Plants 17:1–8

Gan D, Zhang J, Jiang H, Jiang T, Zhu S, Cheng B (2010) Bacterially expressed dsRNA protects maize against SCMV infection. Plant Cell Rep 29:1261–1268

Giraldo JP, Landry MP, Faltermeier SM, McNicholas TP, Iverson NM, Boghossian AA, Reuel NF, Hilmer AJ, Sen F, Brew JA (2014) Plant nanobionics approach to augment photosynthesis and biochemical sensing. Nat Mater 13:400–408

Gordon KH, Waterhouse PM (2007) RNAi for insect-proof plants. Nat Biotechnol 25:1231–1232

Hao R, Xing R, Xu Z, Hou Y, Gao S, Sun S (2010) Synthesis, functionalization, and biomedical applications of multifunctional magnetic nanoparticles. Adv Mater 22:2729–2742

Jiang L, Ding L, He B, Shen J, Xu Z, Yin M, Zhang X (2014) Systemic gene silencing in plants triggered by fluorescent nanoparticle-delivered double-stranded RNA. Nanoscale 6:9965–9969

Kanasty R, Dorkin JR, Vegas A, Anderson D (2013) Delivery materials for siRNA therapeutics. Nat Mater 12:967–977

Kumlehn J, Serazetdinova L, Hensel G, Becker D, Loerz H (2006) Genetic transformation of barley (*Hordeum vulgare* L.) via infection of androgenetic pollen cultures with *Agrobacterium tumefaciens*. Plant Biotechnol J 4:251–261

Ladewig K, Xu ZP, Lu GQ (2009) Layered double hydroxide nanoparticles in gene and drug delivery. Expert Opin Drug Deliv 6:907–922

Lai C-Y, Trewyn BG, Jeftinija DM, Jeftinija K, Xu S, Jeftinija S, Lin VS-Y (2003) A mesoporous silica nanosphere-based carrier system with chemically removable CdS nanoparticle caps for stimuli-responsive controlled release of neurotransmitters and drug molecules. J Am Chem Soc 125:4451–4459

Lau S, Mazumdar P, Hee T, Song A, Othman R, Harikrishna J (2014) Crude extracts of bacterially-expressed dsRNA protect orchid plants against Cymbidium mosaic virus during transplantation from *in vitro* culture. J Hortic Sci Biotechnol 89:569–576

Lau S-E, Schwarzacher T, Othman RY, Harikrishna JA (2015) dsRNA silencing of an R2R3-MYB transcription factor affects flower cell shape in a Dendrobium hybrid. BMC Plant Biol 15:194

Lew TTS, Wong MH, Kwak SY, Sinclair R, Koman VB, Strano MS (2018) Rational design principles for the transport and subcellular distribution of nanomaterials into plant protoplasts. Small 14:1802086

Li H, Guan R, Guo H, Miao X (2015) New insights into an RNAi approach for plant defence against piercing-sucking and stem-borer insect pests. Plant Cell Environ 38:2277–2285

Lübbe AS, Bergemann C, Riess H, Schriever F, Reichardt P, Possinger K, Matthias M, Dörken B, Herrmann F, Gürtler R (1996) Clinical experiences with magnetic drug targeting: a phase I study with 4′-epidoxorubicin in 14 patients with advanced solid tumors. Cancer Res 56:4686–4693

Luo D, Saltzman WM (2000) Synthetic DNA delivery systems. Nat Biotechnol 18:33–37

Mitter N, Worrall EA, Robinson KE, Li P, Jain RG, Taochy C, Fletcher SJ, Carroll BJ, Lu GM, Xu ZP (2017) Clay nanosheets for topical delivery of RNAi for sustained protection against plant viruses. Nat Plants 3:16207

Ohta Y (1986) High-efficiency genetic transformation of maize by a mixture of pollen and exogenous DNA. Proc Natl Acad Sci U S A 83:715–719

Plank C, Zelphati O, Mykhaylyk O (2011) Magnetically enhanced nucleic acid delivery. Ten years of magnetofection—progress and prospects. Adv Drug Deliv Rev 63:1300–1331

Potrykus I (1990) Gene transfer to cereals: an assessment. Bio/Technology 8:535

Radu DR, Lai C-Y, Jeftinija K, Rowe EW, Jeftinija S, Lin VS-Y (2004) A polyamidoamine dendrimer-capped mesoporous silica nanosphere-based gene transfection reagent. J Am Chem Soc 126:13216–13217

Robinson KE, Worrall EA, Mitter N (2014) Double stranded RNA expression and its topical application for non-transgenic resistance to plant viruses. J Plant Biochem Biotechnol 23:231–237

Sanford J, Skubik K, Reisch B (1985) Attempted pollen-mediated plant transformation employing genomic donor DNA. Theor Appl Genet 69:571–574

Schaumberg KA, Antunes MS, Kassaw TK, Xu W, Zalewski CS, Medford JI, Prasad A (2016) Quantitative characterization of genetic parts and circuits for plant synthetic biology. Nat Methods 13:94–100

Scherer F, Anton M, Schillinger U, Henke J, Bergemann C, Krüger A, Gänsbacher B, Plank C (2002) Magnetofection: enhancing and targeting gene delivery by magnetic force *in vitro* and in vivo. Gene Ther 9:102–109

Schreiber D, Dresselhaus T (2003) *In vitro* pollen germination and transient transformation of *Zea mays* and other plant species. Plant Mol Biol Rep 21:319

Sheen J (2001) Signal transduction in maize and Arabidopsis mesophyll protoplasts. Plant Physiol 127:1466–1475

Shukla R, Bansal V, Chaudhary M, Basu A, Bhonde RR, Sastry M (2005) Biocompatibility of gold nanoparticles and their endocytotic fate inside the cellular compartment: a microscopic overview. Langmuir 21:10644–10654

Silva AT, Nguyen A, Ye C, Verchot J, Moon JH (2010) Conjugated polymer nanoparticles for effective siRNA delivery to tobacco BY-2 protoplasts. BMC Plant Biol 10:291

Stöger E, Moreno RMB, Ylstra B, Vicente O, Heberle-Bors E (1992) Comparison of different techniques for gene transfer into mature and immature tobacco pollen. Transgenic Res 1:71–78

Sun D, Hussain HI, Yi Z, Siegele R, Cresswell T, Kong L, Cahill DM (2014) Uptake and cellular distribution, in four plant species, of fluorescently labeled mesoporous silica nanoparticles. Plant Cell Rep 33:1389–1402

Tenllado F, Dıaz-Ruız J (2001) Double-stranded RNA-mediated interference with plant virus infection. J Virol 75:12288–12297

Tenllado F, Llave C, Dı́az-Ruı́z JR (2004) RNA interference as a new biotechnological tool for the control of virus diseases in plants. Virus Res 102:85–96

Torney F, Trewyn BG, Lin VS-Y, Wang K (2007) Mesoporous silica nanoparticles deliver DNA and chemicals into plants. Nat Nanotechnol 2:295–300

Touraev A, Stöger E, Voronin V, Heberle-Bors E (1997) Plant male germ line transformation. Plant J 12:949–956

Wang H, Koleilat GI, Liu P, Jiménez-Osés G, Lai Y-C, Vosgueritchian M, Fang Y, Park S, Houk KN, Bao Z (2014) High-yield sorting of small-diameter carbon nanotubes for solar cells and transistors. ACS Nano 8:2609–2617

Wittrup A, Ai A, Liu X, Hamar P, Trifonova R, Charisse K, Manoharan M, Kirchhausen T, Lieberman J (2015) Visualizing lipid-formulated siRNA release from endosomes and target gene knockdown. Nat Biotechnol 33:870–876

Wong MH, Misra RP, Giraldo JP, Kwak S-Y, Son Y, Landry MP, Swan JW, Blankschtein D, Strano MS (2016) Lipid exchange envelope penetration (LEEP) of nanoparticles for plant engineering: a universal localization mechanism. Nano Lett 16:1161–1172

Wu Y, Phillips JA, Liu H, Yang R, Tan W (2008) Carbon nanotubes protect DNA strands during cellular delivery. ACS Nano 2:2023–2028

Xu ZP, Stevenson GS, Lu C-Q, Lu GQ, Bartlett PF, Gray PP (2006) Stable suspension of layered double hydroxide nanoparticles in aqueous solution. J Am Chem Soc 128:36–37

Zhao X, Cui H, Chen W, Wang Y, Cui B, Sun C, Meng Z, Liu G (2014) Morphology, structure and function characterization of PEI modified magnetic nanoparticles gene delivery system. PLoS One 9:e98919

Zuo J, Niu QW, Chua NH (2000) An estrogen receptor-based transactivator XVE mediates highly inducible gene expression in transgenic plants. Plant J 24:265–273

Chapter 7
Nanotechnology and Plant Disease Diagnosis and Management

Afifa Younas, Zubaida Yousaf, Madiha Rashid, Nadia Riaz, Sajid Fiaz, Arusa Aftab, and Shiwen Haung

7.1 Introduction

Plant infection is a major alarming situation in sustainable crop production affecting 20–30% of the total annual loss (Nezhad 2014). Nutritional sustainability has declared as one of the most crucial concerns of human race. Nation-states, societies and administrations have been fraught with this problem from a long time ago. The imminent appearances even miserable with food deficiency matter approaching hefty. Plant infections also lead towards economic loss by communal attack of plant pathogens on other cash crops than food cash crops resulting in loss in achievable yield (Pan et al. 2010; Thind 2012). World population is increasing day by day therefore the main task exactly is how to fulfill the requirement of growing population with limited resources (less input expense, limited land) and to reduce the dangers to the environment. Main task, sometimes, appears to be disease management.

Nanotechnology is a versatile discipline which embraces information from natural science, chemistry, physics and other fields. Joseph and Morrison (2006) explained nanotechnology as the operation or assemblies of discrete atoms, mole-

A. Younas
Department of Botany, Lahore College for Women University, Lahore, Pakistan

State key laboratory of Rice Biology, China National Rice Research Institute, Hangzhou, China

Z. Yousaf (✉) · M. Rashid · N. Riaz · A. Aftab
Department of Botany, Lahore College for Women University, Lahore, Pakistan

S. Fiaz
Department of Plant Breeding and Genetics, University of Haripur, Haripur, Pakistan

S. Haung (✉)
State key laboratory of Rice Biology, China National Rice Research Institute, Hangzhou, China
e-mail: huangshiwen@cass.cn

S. Javad (ed.), *Nanoagronomy*, https://doi.org/10.1007/978-3-030-41275-3_7

cule or molecular collections into structures to generate material maneuvers with novel or extremely diverse assets. The utilization of nanotechnology in agriculture can modify the agricultural science with advanced apparatuses for quick infection recognition, directed dealing, improve the capability of vegetation to engross nutrients, combat contaminations and resist ecological stresses and active operational schemes for handling etc. Smart sensors and smart delivery systems will support the agronomic production to fight against viruses and further harvest infections.

Nanotechnology has another prospective to improve the agronomic harvest from genetic amendments of vegetation, transfer of gene and pesticide to exact sites at cellular level. Nano array and genomic technologies can be used for strengthening the plant defense system in disease condition or stress management.

Initial and effective detection of pathogen is vital for proper infection resistance and management to reduce the yield damage. Different types of nanoparticles like gold, magnetic nanoparticles and quantum dots are generally used for molecular level detection. Due to some distinctive properties of gold nanoparticles like small size, catalytic and surface effects, these are used for quick insusceptible identification. Gold nanoparticles can show covalent bonding with DNA so it can be used for DNA assessment and reorganization. Nano Plant Pathology is a progressive branch of science in which nano based applications are used for detection, diagnosis and controlling of plant pathogens at preliminary phase, owing to harvest protection from endemic ailments.

New phyto pathologists are trying to use their information in nano plant pathology for diagnosing and monitoring features of plants illness and to assess environmental friendly analytical dimensions. Recent nano based molecular techniques are used for checking the microbes populace inheritances, gene transmission and plant microorganism interactions between host and pathogen. Moreover, sliver and silica nanoparticles have been used as antifungal and antibacterial means. Furthermore, nanoparticles can be used for fungal toxin diagnosis and detoxification, improving host resistance (plant) and nano-based prediction of phyto-pathogens.

Bio-barcode analysis is an ultra-sensitive magnification, acceleration and recognition of nucleic acids/proteins. DNA bio-bar coded examinations can be used for oligonucleotide-modified magnetic gold nanoparticles (AuMNPs), indication of detected magnification and mild parting of a specific protein from the sample. The colossal amount of DNA can be detected by identification mediator which manages a means of significant indication/recognition. Different types of proteins can also be detected by its molar concentrations (Joseph and Morrison 2006) and nucleic acids can be identified by their molar levels under the improved conditions (Goluch et al. 2006). The conception of the bio-barcode evaluation is exclusive and effective which characterizes a latent substitute to the PCR practice. This chapter describes about uses of nanotechnology for faster, more gainful and accurate analytical measures of plant ailments. Nanotechnology might have main effects on flora in future. Coped or skillful usage of the nanotechnology might create chances for evolving novel procedures which will be quicker, accurate and effective diagnostic systems (Nam et al. 2004).

7.2 Nanotechnology Based Detection of Plant Diseases

In plant disease management, one of the serious issues is the detection of right phase of disease. A large proportion of plant disorders are seen at later stages only thus their control becomes a challenging task. Mostly, pesticides and fungicides are applied on plants for protection, only after the appearance of symptoms. Instead of plant protection, these cause significant loss of crops. Subsequently, it is critical to distinguish and dissect plant diseases at their starting stage, with the objective that the plant defensive synthetic chemicals could be given at right amount and at ideal time, so as to stay away from residual toxicity and environmental hazard. A shrewd coordination among nanotechnology and plant pathology could help in the early location of various viral, parasitic, and bacterial sicknesses in plants. Existing recognition procedures ordinarily take numerous days to recognize the illness, in this way scientists more often than not focus on straightforward identification strategies that give better results within a little time period. There are number of problems associated with disease detection in field plants. Sometimes, it is not possible to take sample to lab for diagnosis. Furthermore when a crop is established in a field, it is not possible to check each and every plant in the field for disease symptoms. These kinds of frameworks are specific and in this manner for the most part fitting for research facility utilize as it were. In reasonable cultivating, distinct data of the scattering of infections in the field is significant; however it is difficult to look at each and every plant in a tremendous field an area. Additionally, it is troublesome, repetitive and exorbitant also.

Appropriate identifying frameworks that could perceive and assess pathogens in particular places of the field would be really valuable to the ranchers in site-focused and expanded utilization of agrochemicals alongside least ecological dangers. In this circumstance, nano-biosensors, once presented in the field, could perceive pathogens with extraordinary particularity and affectability. These nano-sensors are very compact frameworks with 'real time' checking of results. They don't require complex sample arrangements and furthermore the location is targeted (Sadanandom and Napier 2010). Cell biologist of Cornell University have been taking a shot at nanofabrication advances to perceive how the microorganisms feel their way on the plant surface, to actuate tainting (Mccandless 2011). Fungi can recognize little differentiations on the leaf surface to pick the spot and time to sully. Scientists have recreated leaf surface by creating smaller scale edges on silicon wafers utilizing electron beam lithography and organisms are made to situate themselves on edges, which are fundamentally the same feel as the leaf surface. This will help to understand the pest attack and mode of action. Along these lines the information could support the reproducers (in not so distant future) to develop better assortments that makes them significantly increasingly impervious to contagious attack. Viral infections are very difficult to control, among all plant diseases, as in this case, the vectors which are spreading the diseases needs to be prevented instead. Also, the indications of disease appear very late, therefore, the application of pesticide after the advent of symptoms is not so much successful. Along these lines, recognition of exact phases of disease either viral DNA replication or viral protein production is very important to cure such infections. Utilization of nano-based viral indicators could help in the recognition of particular viral strain as well as the exact phase for the application of protective chemicals to control these diseases. Such nano-based diagnostic kits would be able to detect the pathogen and disease at a quicker speed and also lead towards the enhanced power of detection.

7.3 Nano-phytopathology

Various sorts of inorganic and organic salts have been utilized for controlling infection for a longer time (Talibi et al. 2011). There are a number of reports by various scientists to use different amendments for a better and enhanced control of plant diseases (Abd El-Hai et al. 2010; Kundu and Saha 2014; Ali et al. 2017). Citrus blue mold was controlled by the organic acids and salts (Latifa et al. 2011). Nanotechnology is a novel field for this part of century, but with passage of time and continued research in the field, utilization of nanoparticles in treatment of plant infection will be increased. Nanotechnology is going to offer a bright future of plant pathology studies in reference to their antibacterial and antifungal action. The most effective way may be to shield the seeds and foliage from intruding pathogens by use of nanoparticles. Along these lines, the NPs may stifle the pathogens in a manner practically identical to synthetic pesticides. But when we talk about broadly, about application of nanoparticles on soil and their interaction with non-targeted species should be considered. Particularly when these non-targeted species are playing their beneficial roles in soils, like nitrogen fixing bacteria, then this problem becomes really serious. Research also shows that nanomaterials can be used as a carrier of important chemicals like pheromones, inhibitors, pesticides, nutrients etc. Also, the nanomaterials, carbon tubes, cups and so forth can likewise be utilized as a bearer of some extravagant synthetic substances for example, pheromones, SAR actuating synthetic substances, polyamine combination inhibitors or even focused dynamic elements of pesticides for their controlled discharge. Therefore it can be summarized here that nanoparticles can play their role in plant disease management in following ways mainly,

- Nanoparticles being used as pesticides themselves and being applied to plants directly for the control of disease
- Nanoparticles as carriers of other pesticides/nanoparticles for their controlled and targeted release and to increase their effect
- Nanodevices to detect diseases at early stages

Nanoparticles can be successfully applied to disease affected plants with enhanced and effective results due to their extremely reactivity/affectability. This increased reactivity can be attributed to their extremely small size and large surface area.

7.3.1 *Effect of Nanoparticles on the Pathogens/ Microorganisms*

Since, chemical and physical properties of nanoforms of materials have shown really marvelous shift from its macroform to nanoform. These transformations in properties end up with real practical applications in plant protection and plant defense. Nanoparticles influence the activity of plant pathogens in a more precise way due to their advantageous size and surface area. They have more chances of interactions with microorganisms as compared to their macro sizes. There is a lot of research available to strengthen this argument.

7.3.2 *Effect of Nanoparticles on Bacteria*

Recent studies have shown that nanoparticles can have antibacterial properties, caused either by physical destruction of the bacterial cell wall or by oxidative stress through generation of reactive oxygen species (ROS) (Zhang et al. 2010; Cabiscol et al. 2007; Wiesner et al. 2006).

Bacterial infection is leading cause of prolonged contaminations and mortality. Antibiotics have been chosen for the treatment of bacterial contamination because of low cost and effective outcomes. But several studies have indicated that the overuse of antibiotics leads towards the multidrug resistant bacterial strains. Some kind of super bacteria have developed resistance to almost all kind of antibiotics. Previous researches have exposed, these kinds of bacteria consist of gene, which are super resistant (Hsueh 2010). Nanoparticles can control these types of super-resistant bacteria because mode of action of nanoparticles is directly linked with bacterial cell wall. Nanoparticles would be less susceptible to promoting resistance of bacteria than antibiotics. So, it would be possible that new NP-based materials can perform antibacterial activity (Wang et al. 2017a, b, c). Latest researches have shown that nanoparticles contain anti-bacterial properties, which showed either physical destruction of bacterial cell wall and resultant formation of reactive oxygen species (ROS) which causes oxidative stress (Gurunathan et al. 2012).

Silver nanoparticles, Zinc oxide nanoparticles, Copper nanoparticles have been reported to have enhanced antimicrobial activities against *Staphylococcus aureus*, *Escherichia coli*, *Pseudomonas aeruginosa* and *Klebsiella pneumonia* etc. Nanoparticles may be bactericidal or bacteriostatic but mainly nanoparticles inhibit their growth by controlling the metabolism or by decreasing their colonization. There are a lot of examples from the literature which support the results like maximum inhibition value against the species *P. aeruginosa* was shown by zinc nanoparticles (Jayaseelan et al. 2012). Antibacterial activity of hybrid material which were based on nano-particles of silver and polyvinylpyrrolidone was shown for three kinds of bacteria, gram positive bacteria (*Staphylococcus aureus*), non-ferment gram-negative bacteria (*P. aeruginosa*) and gram-negative bacteria (*E. coli*) studied by Bryaskova et al. (2011). Depending on NPs concentration silver nano-particles affect the bacterial physiology, inter-cellular permeability of membranes and metabolism of microbes. The nanoparticles showed high anti-microbial activity against *P. aureginosa*, *S. aureus* and *E. coli* (Guzman et al. 2009). There is another nanomaterial copper oxide nano-particles (CuO NPs) disclosed antibacterial activity against *Bacillus subtilis*, *E. coli*, *S. aureus* and *P. aeruginosa* (Azam et al. 2012).

7.3.3 Influence of Nano-particles on Plant Pathogenic Fungi

A fungal disease can be identified by symptoms which are produced on the particular part of the plant. Diagnosing and controlling plant pathogenic microbes (fungi, bacteria, virus and nematodes) which cause the plant diseases is the major task for nanotechnologist. These microbes are significant restrictive reasons for progression of crops and foodstuffs (Khan et al. 2011; Khan 2012).

Different methods and techniques are being functioned to diagnose pathogen but there is no proper way for control of infection (Khan and Jairajpuri 2010a, b; Khan 2012). Most of the nanoparticles have a great scope for management of plant pests and nanoparticles area mostly effective to control the fungi. Nano-forms of micronutrients ($CuSO_4$ and $Na_2B_4O_7$) exhibited the maximum operative to control rust disease of *Pisum sativum*. Micro-nutrients such as Mn (manganese) and Zn (zinc) particles inhibited the damping off and charcoal rot diseases in *Helianthus annuus* (Abd El-Hai et al. 2010).

The Silver nanoparticles and Polyvinylpyrrolidone (AgNPs/PVP) show their own fungicidal activity in counter to various molds and yeasts such as *Candida krusei*, *C. albicans*, *C. glabrata*, *C. tropicalis* and *Aspergillus brasiliensis*. The mixture of PVP and Ag nanoparticles exhibited sound anti-fungal effects against microorganisms (Bryaskova et al. 2011). Zinc-nanoparticles with fungicides worked against two patho-fungi *Botrytis cinerea* and *Penicillium expansum*. Anti-fungal activity of ZnO NPs can be inspected by Scanning Electron Microscopy (SEM) and Raman Spectroscopy to highlight any changes in cellular and morphological composition of fungal hyphae. The NP handlings produced alteration in the hyphae of *B. cinerea* and prohibited the asexual growth of *P. expansum* which ultimately became reason of losing fungal propagation (He et al. 2011).

Effects of silver nanoparticles were also examined on plant patho-fungi, *Macrophomina phaseolina*, *Sclerotinia sclerotiorum*, *Alternaria alternata*, *B. cinerea*, *Curvularia lunata* and *Rhizoctonia solani*. The lesser concentration of NPs exhibited more inhibitory activity of all the tested pathogens (Krishnaraj et al. 2012). The Zn NPs (25 mg mL^{-1}) can also terminate the growth of *A. flavus* (Jayaseelan et al. 2012). These nanoparticles have been found effective antifungal agents which cause positive consequences than synthesized fungicides.

7.4 Bio-nano Materials

The bio-nano materials are manufactured by bio reduction methods in which simple sugars, catalyst proteins and other complex phenolic compositions are suggested to cause the reduction. The manufacturing process has been monitored by different factors like reaction mixture (pH), origin of material and material nature. After formation of the bio-nano materials are categorized through X-ray diffraction (XRD) technique, X-ray photoelectron spectroscopy (XPS), Energy-dispersive X-ray spectroscopy (EDS), UV visible spectroscopy, Scanning electron microscopy (SEM), Fourier transforms infrared spectroscopy (FTIR), Coupled plasma spectrometry (ICP), Transmissions electron microscope (TEM) and Atomic force microscopy (AFM) techniques.

Metal nano particles have outstanding uses in the arena of life sciences like catalysts, sensors, optically used thin films, diagnostic probes and other displayed devices. These bio nano materials played noteworthy role in field of agriculture, medicine and biology. Even these nano-materials are used in removal of pollutants from environment. The increasing usage of bio nano materials in many areas will enhance their productivity in the atmosphere by developing more analytical tools in nanotechnology for controlled environmental risk management (Sahayaraj 2012).

Bio-synthesis of bio-nanomaterials can be achieved by using plant extracts (Sahayaraj et al. 2015), microbial cultures or their enzymes and proteins. They also possessed anti-feedant, larval killing by formulating with silver nanoparticles using liquid leaf extract of *Aristolochia indica* against growing stage of larvae of *H. armigera* which revealed maximum anti-feedant and larvae killing efficiency (Siva and Kumar 2015). Gold nanoparticles and bifenthrin insecticide showed more intense effect to *Lygushesperus*. Another combination of bifenthrin and Ag solution was toxic against *Acheta domesticus* treated eggs under cotton filed condition due to arrested embryonic development (Louder 2015).

7.5 Nano Bio-barcode Assay

Bio-barcode is a developing technique with advancements in nanotechnology used for recognition of enzyme-free ultra-sensitive proteins and DNAs. Protein bar-code assay would be more complex, more sensitive and profound than orthodox ELISA based assays relying upon target and sample density. The nanoparticle based bio-barcode assay is very sensitive towards detection of pathogens in comparison to other traditional techniques like ELISA, Real time PCR etc (Bao et al. 2006; Nam et al. 2003). This can help to detect plant disease at a very early stage. The bio-barcode technique consists of two probes

1. Magnetic micro beads (MMB): which are meant for target recognition and carry an antibody or DNA as a biological probe

2. Gold nanoparticles (Au-NP): it has a polyclonal antibody or an oligonucleotide (Bio-barcode)

Bio-barcode is a developing technique with the help of advancements in nanotechnology. It is an enzyme free, PCR free technique and highly sensitive for protein and DNA detection.

Traditional conventional approaches seems unable for handling complex level probability existing biomarkers while bio-barcode method was helpful in diagnosis of low concentration biomolecules with prostate specific antigen (PSA).

One of the example in human beings is with prostate specific antigen (PSA) test. Nowadays, advanced bio barcode PSA assay is found to have 300 times additional sensitivity facilitating measurement of serum PSA level comparable to commercial immunoassays. Hence, this advance technique is helpful in significant possible outcomes i.e. (1) Detection of non-rising PSA level which was undetectable with conventional assays, (2) Diagnosis ability to assign earlier recurrence by measuring increased PSA levels which was impossible through conventional assays (3) To use PSA levels to find out patients response towards their recovery therapies (Bao et al. 2006).

Advancement has been made on different aspects of Bio-barcodes to make it more applicable in fields. DNA barcoding has been suggested for fungal identification (Xu 2016). It has been reported to be reliable and rapid method of detection. A DNA barcode should be standardized and scalable. It should be a protein coding region, easy to make copies and should be easily comparable.

Fluorescence bio-barcode technology is used to identify *P. aeroginosa* with gold nanoparticles. In probe 1 bio barcode DNA who act as signal identifier, second probe was designed for identification of specific DNA at other end. So, both probes worked and hybridized to their complementary DNA sequences through covalent bonding. The fluorescent spectroscopy proves the assay is fast, convenient, sensitive and has a wide linear range of (5–200 ng/mL) detection (Amini et al. 2017).

Similar techniques can be developed for speedy and onsite detection of plant pathogens particularly viruses to decrease the losses to crops.

7.6 Nanopore System

Nano pore systems are capable of electronic detection of DNA structures with low cost, less sample preparation potential and working at high speed (Branton et al. 2008). In fact nanopore is a pore of nano size and ionic current is passed through the pore. A change in that current is measured as that informs about biological molecule. Nanopore based systems identify nucleotides by measuring conductivity changes which can detect nucleotide through lipid membrane whereas, DNA section is dragged across a nano-scale pore by electric current (Egan et al. 2012). Protein nanopore is injected in a polymer bilayer membrane from the top of a micro well which bear sensory chip to measure iconic current (Clarke et al. 2009).

Newly, nanopore based sequencing (Nano-SBS) distinguished four DNA bases through discovering four different sized tags released from 5′-phosphate-modified nucleotides at the particular molecule level for sequence determination (Kumar et al. 2012). Nano SBS can sequence the real long frame of DNA and RNA. On the other hand IBM and Roche introduced an advance sequencing technology "DNA transistor" having the potential to record the nucleotide sequence by pulling template through nanopore sensor (Zhang et al. 2011). Currently, UK based nanopore technologies has launched a portable DNA sequencing machine (MinION) which enables to sequence 10 kb of single sense and anti-sense DNA strand to make next generation sequencing easily approachable to other researchers (Hayden 2015). This advance technology offers not only determination of epidemic outbreak, epidemiology of disease but can discriminate among closely related microorganisms i.e. bacteria, fungi, viruses, complex genomic portions and difference among two versions of genes located on individual chromosome. Thus, nanopore platform applicable inside current diagnostic apparatus can analyze the entire genomes in minutes. With reference to agriculture, this technology could be used for analyzing plant and pathogen genomics for improvement of agronomic crops.

7.7 Nanodiagonastic Kit

Nanodiagonastic kit also called "lab in a box" is used as a small box for measuring important tasks in plants which can be done in small space (Khiyami et al. 2014). A smart kit help to detect the plant pathogens and can help the farmers in prevention of wide spread diseases (Pimentel 2009; Nezhad 2014). Nanodiagonastic kit contained four myco-sensors which can detect the of ZEA, T-2/HT-2, DON and FB1/FB2 myco-toxins on only one strip used for cash crops like wheat, barley and corn (Lattanzio et al. 2012). This technique is quick, easier and less expensive for detection of fungal attack on cash crops. There are numerous other purposes, efficient, specific about antigen and antibodies, nucleotide sequence in which nano kit can be used. So, it required more improved features for different applied agriculture sectors. It required evaluations of different trials and specific level for detection limits. Moreover, it can also detect particular gene target, isolation and purification of specific genes. But nano kit has not fully checked practically for the plant pathogen detection in field conditions. More extensive research work is needed in this field.

7.8 Quantum Dot (QDs)

Quantum Dot (QD) is another level of radiant semiconductor nanocrystals that radiate light of specific wavelengths. They are three dimensional nanoparticles (Edmundson et al. 2015) offering many benefits due to their wide excitation spectra. These QD possess narrow tunable emission peak, extensive fluorescence lifetime, resistance against photo bleaching and 10–100 times higher molar extinction coefficient. They can assist the fluorescent dyes to result in brighter probes as compared to traditional fluorophores (Zhao and Zeng 2015). Quantum dots are supposed to be the future of indoor planting due to their ability to produce light of particular wavelength.

Quantum dots, also called inorganic fluorophores, propose substantial benefits above conventionally used fluorescent markers. These are highly sensitive (brighter imaging signals) and possess vast excitation bands, with consistent simply excited fluorescence independent of laser. Above stated qualities lead towards QD-FRET-based nanosensors which got universal popularity in agriculture with frequent applications in DNA and enzymatic activity detection region (Stanisavljevic et al. 2015). Detection of plant viruses is very important to secure the future. It's of critical importance when you are concerned about international trade of food and crops etc. QD related techniques when came forward to play their role as they are rapid and cost effective as compared to ELISA and PCR.

Knudsen et al. (2013) in one of his findings revealed multiple enzymatic properties of QD-based nanosensors and Safarpour et al. (2012) also stated CdTe quantum dots usage as biosensors with specific antibiotic coatings against *Polymyxa betae* specific glutathione-S-transferase (GST) protein. The mutual antigen-antibody bonding carrying combination of CdTe quantum dots and rhodamine permit the resonance dipole-dipole coupler needed for the occurrence of the fluorescence resonance energy transfer (FRET). The assembled immune sensor exhibit highly sensitive, specified and effective utilization for maximum output to screen plant samples in less than 30 min time. Contrary to this Rad et al. (2012) documented a highly sensitive detection of phytoplasma (*Candidatus Phytoplasma aurantifolia*) diseased lime trees by showing 100% specificity with a sensitive detection limit for *P. aurantifolia* successfully by quantum dot (QD) based nano-biosensors.

Currently, an optical fluorescence resonance energy based upon DNA biosensor with synthesized quantum dot has been recognized for the detection of specific-sequence of DNA for *Ganoderma boninense* (Bakhori et al. 2013). Synthesized quantum dots (5–8 nm) comprised of carboxylic groups and conjugated with a single-stranded DNA probe (ssDNA) via amide-linkage when hybridized with target DNA conjugated with QD-ssDNA along with reporter probe labeled with Cyanin 5 allows the detection of associated synthetic DNA sequence of *Ganoderma boninense* gene based on FRET signals. The newly established biosensors were reported to have sensitivity detection limit of 3.55×10^{-9} M providing the proficient, simple, fast and sensitive method of plant pathogens detection. Moreover, QDs excitation with ultraviolet (UV) light and fluorescence is possible to visualize with bare eyes and can be applicable into field for instant use. Research is the utmost necessity of beginning of QDs to check their efficacy to detect plant pathogens and toxic compounds in food. Still, there is much of the work needed for the optimization of assays to get precise signals to identify minor stages of pathogens with multifaceted systems either food, plants or insects etc. It provides unlimited opportunities for QDs nanoparticles constructed diagnostic application in agricultural setups to keep in touch with advanced technologies for betterment of crops. It can create a revolution in Agronomy.

There are reports of production of quantum dots by fungi and other microbes. The mycosynthesis of semiconductor nanomaterial response against salt stress was first detected in unicellular yeast by turning them out into Cadmium sulphide crystals as a result of cadmium salt stress (Cds) (Dameron et al. 1989). Diverse microorganisms have been reported to be useful for synthesis of cadmium quantum dots but its fluorescent properties have been remained limited (Yadav et al. 2015). Whereas significant myco-mediated biosynthesis of fluorescent CdTe quantum dots was also achieved by *Fusarium oxysporum*, when mixed with cadmium dichloride ($CdCl_2$) and Tritellurium dichloride ($TeCl_2$) (Jain 2003; Kashyap et al. 2013; Alghuthaymi et al. 2015) (Fig. 7.1).

Quantum dots have a lot of applications and already in final steps in medicine and diagnostics for human health. For example, QDs with infrared coloration allows complete blood analyses with broad ranged applications of molecular diagnostics and genotyping. They also contribute in complex diagnosis and combination with therapeutics leading towards possible applications for cancer diagnosis. Some specific quantum dots like luminescent and stable QD bio conjugates enables to visualize cancer cells in living animals as well as provide high resolution of cancer cells in combination with fluorescence microscope.

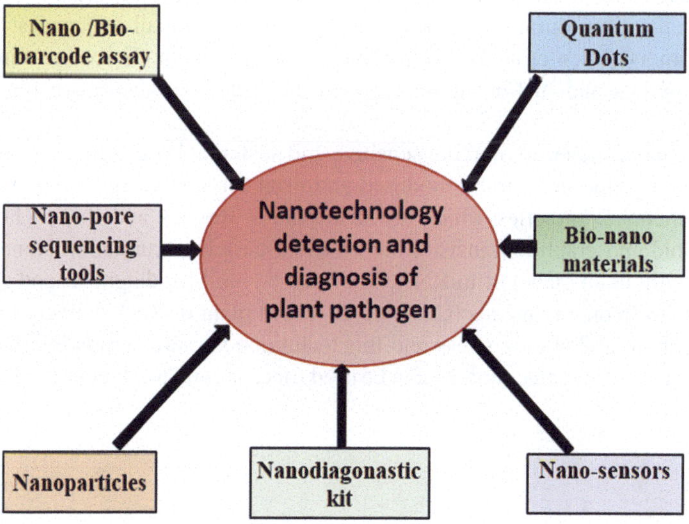

Fig. 7.1 Effect of different nanotechnology approaches for detection and diagnosis of plant pathogens

7.9 Nanotechnology and microRNA for Harvest Security

RNA molecules have clear and attracted attention due to their remarkable range in construction and task. RNA can be in numerous forms like ribozymes, aptamers, miRNA (micro RNAs) and siRNA (small interfering RNA) (Shu et al. 2013). The RNA nanotechnology is a unique method for dealing and identification of several types of tumors, viral contagions and genetic infections in plants (Chaudhary et al. 2018).

The non-coding miRNA is endogenous which is mainly responsible for gene expression in living organism. These miRNAs commonly has 18–22 nucleotides and these are most important in translational regulation. These miRNAs are active and dynamic for their functions and produced by nucleus. The miRNA configuration is altered in disease stated miRNAs, a lavish task for medication remedy through RNA nanotechnology. Renewal of inactivated miRNA or highly-expressed miRNA to return miRNA to its normal condition is the evidence for infection control which can be moreover examined to analyze, control and manage further several protein/enzymes based functions also. There are specialized genes for harvest protection which can be maintained by the handling of miRNAs. Plant miRNAs (extremely flexible in size) are commonly transcribed by RNA pol II as alike to animals (Axtell et al. 2011; Ma et al. 2015).

miRNAs are the main key managers of changed gene product due to attack of diseases in living organisms (plants, animals and humans). So, accurate and sensitive evaluation of miRNA can correct disease diagnosis at early stage and these can act as biomarkers for further research (Degliangeli et al. 2014a, b). In previous research infected miRNA detective method was based no microarrays and real time polymerase chain reaction. These techniques are erring through hybridization and need specific time duration because miRNA are very small sequences; difficult coded primers (Chen et al. 2005). Nanotechnology is right choice for minimizing the ambiguities and making more targeted and sensitive alternative for miRNA detection.

Gold nanoparticles are making sensitive and sustainable detection technology as their optical properties can be modified and used for surface functions with well-defined chemical moieties which are disulphides, amines and thiols (Jain et al. 2007). Gold and graphite transistor biosensors are more accurate, efficient and can be used without any label of miRNA. They can be used for diagnosis and detection of disease in living organisms (Cai et al. 2015). In plant miRNA delivery technique used for biotic and abiotic stress and this technique is more compatible than other techniques. RNA nanotechnology can be used for crops and food security (Table 7.1).

Table 7.1 Effect of nano-particles mRNA and plant disease

No of Obs.	Nano-particles	mRNA	Plant disease	Reference
1.	Artificial miRNA in transgenic Arabidopsis	amiR-P69^{159} and amiR-Hc-Pro^{159}	(1) Turnip yellow mosaic virus (2) Turnip mosaic virus	Niu et al. (2006)
2.	AmiRNA based on Arabidopsis	pre miRNA159a	Cucumber mosaic virus	Ai et al. (2011)
3.	AmiRNA against Hc-Pro	*Arabidopsis thaliana* miRNA159a, miRNA167b and miRNA171a	(1) Potato virus Y (PVY) (2) TGBp1/p25 of Potato virus X (PVX)	Duan et al. (2008)
4.	*Nicotiana tabacum*	amiR-159a	Cassava brown streak virus	Wagaba et al. (2016)
5.	*Vitis vinifera*	Grapevine fan leaf virus	amiR^{CP}-2	Jelly et al. (2012)
6.	*Solanum lycopersicum*	Viral infection	amiR-2a/b	Zhang et al. (2011)
7.	*Nicotiana tabacum*	Potato virus Y and potato virus X	amiR-Hc-Pro^{159a} amiR-Hc-Pro^{167b} amiR-Hc-Pro^{171a}	Ai et al. (2011)
8.	*Arabidopsis thaliana*	Cucumber mosaic virus	amiR-159a	Duan et al. (2008)
9.	*Gossipum hirsutum*	Cotton leaf curl virus	Pre-miR-169a	Ali et al. (2013)
10.	*Triticum aestivum*	Wheat streak mosaic virus	Pre-miR395	Fahim et al. (2012)
11.	*Arabidopsis thaliana*	Water melon silver mottle virus	Pre-miR-159a	Kung et al. (2012)
12.	*Solanum lycopersicum*	Tomato leaf curl virus	amiR-AV1-3	Van et al. (2013)
13.	*Arabidopsis thaliana*	Turnip mosaic virus	amiR^{159}-P69	Lin et al. (2009)
14.	*Oryza sativa*	Rice stripe virus and Rice black streaked swarf virus	Osa-pre-miR528	Sun et al. (2016)
15.	*Hordeum vulgare*	Wheat dwarf virus	huv-pre-miR171	Kis et al. (2016)
16.	*Nicotiana tabacum*	Tomato spotted wilt virus	Pre-miR-159a	Mitter et al. (2016)

7.10 Nanoparticles for the Control of Disease and Pest Incidences in Plants

Some nanoparticles have specific activity or characteristics to bind with specific tissue in its active or passive form. This characteristic of nanoparticles can be used plant disease diagnostics directly. These nanoparticles include gold nanoparticles, magnetic nanoparticles and quantum dots.

7.10.1 Gold Nanoparticles for Diagnostics

Now a day, gold nanoparticles are frequently used for diagnosis. For this purpose, tiny DNA segments smaller than 13 nm are attached to these nanoparticles. Then they will bind to the sensor surface by accompanying a complementary target. Furthermore, a patterned surface sensor occupying multiple DNA strands is attached, then millions of DNA sequences can be detected all together because of having a variety of analytical techniques (Fig. 7.2).

7.10.2 Silver Nanoparticles [AgNPs] and Nano-silver-silica Composites

In case of antimicrobial application, silver nanoparticles have proved to be an effective antifungal and antibacterial mean (Panacek et al. 2009; Singh et al. 2008). Their broad spectrum antimicrobial potency has leds them towards acting as controlling agents against bacterial and fungal diseases in plants. In previous researches, AgNPs showed strong efficacy against *sclerotium* fungi phytopathogens like *Rhizoctonia solani*, *Sclerotinia sclerotiorum* and *S. minor* widely affecting plants resulting into major economic losses. To overcome this situation AgNPs even with small concentrations could be used to minimize this issue by their intensive permeability and contact with fungal pathogens inhibiting growth and sclerotial germination. In this regard their efficacy can be enhanced by pre applications before fungal penetration and colonization in plant tissue cells (Jo et al. 2009). In another research efficacy of AgNPs was also tested against ascomycetous fungi i.e. Raffaelea sp. producing oak wilt disease (Kim et al. 2009). AgNPs significantly decrease the fungus hyphae

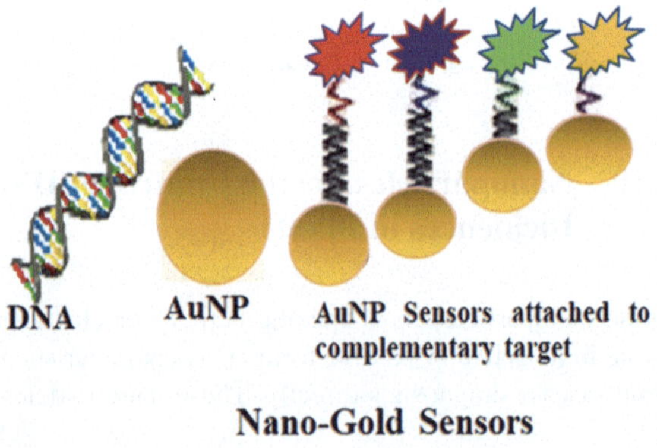

Fig. 7.2 Diagnosis and detection of DNA by Nano-gold sensors

growth and conidia germination in concentration dependent manner. In another case, AgNPs treatments also revealed their intended results against bacterial growth extending vase lifetime of *Gerbera jamesonii* cv. Ruikou flowers (Liu et al. 2009; Solgi et al. 2009).

Nano-silver-silica compound is made up of silver salt with silicate and water soluble polymers and radioactive exposure is reported to possess significant antimicrobial effect with minute quantity (Oh et al. 2006). Therefore, these nano-sized silver silica compounds can be utilized successfully against plant pathogens like *Phytophthora* spp., *Rhizoctonia* spp., *Colletotrichum* spp., Botrytis spp., *Magnaporthe* spp. and *Pythium* spp. (Park et al. 2006). When these nano-composite compounds are absorbed by fungal strains, AgNPs increases sterilizing activity while silica particles physically block pathogenic fungi, thus increasing resistance to disease. Even though, several research evidences declared more effectiveness of nano-composite particles at lower concentrations than 3.0 ppm. While in case of phytopathogenic bacteria, this composition was found effective at concentrations higher than 10 ppm. Therefore, it is essential to optimize the minimal effective nano-formulation for effective microbial defense. The best thing about these composite nanoparticles is their long term control over microorganisms based upon concentration optimization with single application without any toxicity. Such nano-formulation could be proved effective against phytopathogens resistant to antibiotics (Nair and Kumar 2013) (Fig. 7.3).

7.10.3 Nano-carbon

Carbon has its wonderful applications in a number of areas. Being an astounding element, it is available in many forms like diamond, fullerenes, graphite and graphene. Carbon based nanostructures have vast applications for development of each

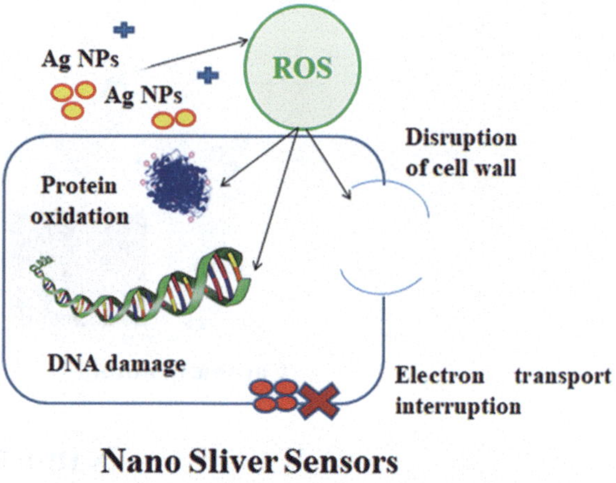

Fig. 7.3 Fungal detection by nano sliver sensors

biological and chemical sensor and their application in agriculture and medicine (Kurbanoglu and Ozkan 2018).

Carbon-based nanomaterials (CBNs), like fullerenes, graphene, carbon nanotubes, nano-diamonds and carbon nanodots, have just received substantial consideration among scientific communities due to their distinctive physio-chemical properties (Hu et al. 2017). These carbon-based nanomaterials stand out as unique nanosensors due to their utmost performance in the detection of heavy metal ions, food additives, gas molecules, toxic pesticides, antibodies and reporters for bio imaging as well. Wang et al. (2017b) stated aerogel preparation from graphene oxide/cellulose nanofibril hybrid for adsorptive removal of four types of antibiotics, achieving 81.5%, 73.9%, 79.1% and 79.5% removal percentages for doxycycline, tetracycline, oxytetracycline and chlortetracycline respectively. Xiong et al. (2018) reported the preparation of $BiVO_4$-rGO (a leaf-like $BiVO_4$-reduced graphene oxide composite) that showed comparatively higher photocatalytic capacity of $BiVO_4$-rGO towards rhodamine B dye degradation under visible-light irradiation than that of pure $BiVO_4$. Similarly, Younes et al. (2017) studied the electrical responses of carbon nanotubes-epoxy (nanocomposites) which were single walled due to the impact of saline solution. They found 50% increased resistance on the nanocomposites film surface by just a drop of saline solution (Fig. 7.4).

7.10.4 *Magnetic Nanoparticles*

The magnetic nanoparticles for site-focused delivery of medications have generally been utilized in the field of biomedicine against a few infections (Jurgons et al. 2006; Mornet et al. 2004). Though, in plant sciences, this type of application is still in its budding stage. However, magnetic nanomaterials could be used for site-targeted delivery of fundamental plant protection compounds as well as against site specific ailments of plants. If it could be probable to remotely follow the movement

Fig. 7.4 Application of nano carbon sensors

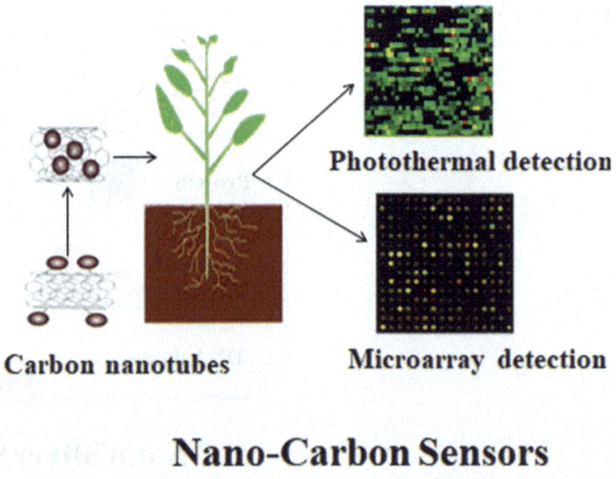

Fig. 7.5 Application of nano magnetic sensors

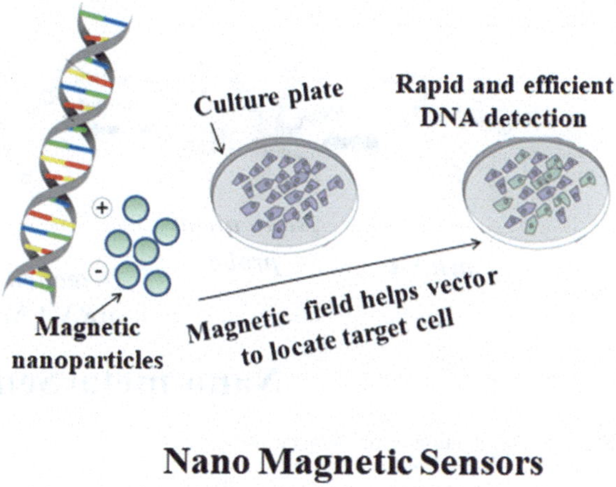

of internal magnetic nanomaterials by utilizing high power external magnets, only then the release of chemicals will become site specific where needed.

Along these lines the functionalized magnetic nanoparticles will permit outer control of the development of nano-carriers inside a plant framework. The nanoparticles should have been functionalized remotely so as to make them progressively biocompatible. Since the presence of these particles inside the plant framework may cause changes in various enzymatic and metabolic elements of plants. This has shown effects on the plant pigments and DNA (Răcuciu and Miclaus 2007, 2009), therefore, further research is required to comprehend the overall changes occurring in plants based on their physiological and metabolic responses for the uptake of magnetic nanoparticles (Fig. 7.5).

7.10.5 Other Metal Nanoparticles

Nanoparticles show sharp bias from their mass in various respects which has turned out to be favorable in developing new analytic tools. Many nano-crystals are attractive probes of biological markers because of a lot of properties. These include their small sizes (1–100 nm), larger surface to volume ratio, alteration in the physiochemical properties with reference to their shapes and sizes, chemically adjustable physical properties, strong relation towards the target proteins. These characters cause the enhancement in the aggregation of particles which was dependent on surface modification types, structural strength, enhanced photoemission, high heat and electric conductivity as well as improved surface catalytic activity (Garg et al. 2008; Liu 2006; Rosi and Mirkin 2005; McNeil 2005) (Fig. 7.6).

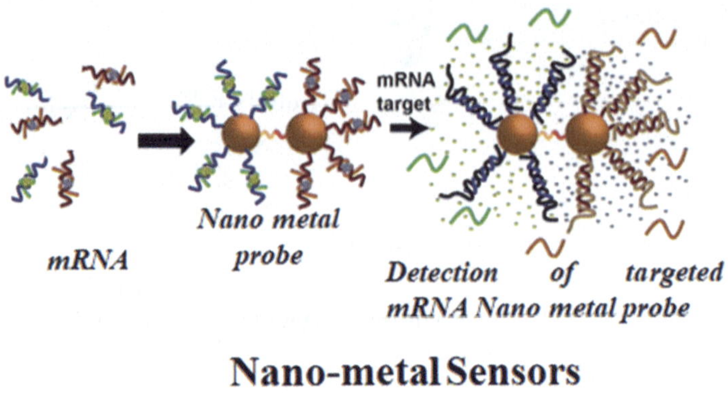

Fig. 7.6 Application of nano metal sensors

7.11 Future Prospects of Nanotechnology in Disease Management

The utilization of molecularly focused on radio-labelled nanoparticles may offer numerous advantages over the recently utilized atomic imaging tests. Primarily, thousands of imaging labels or a combination of labels could be attached to a single nanoparticle, for different imaging modalities. Additionally, various, pesticides may be going to vanish from plant protection protocol. They also seem to be working in pre and post-harvest disease management. Nanoparticle can give improved receptor binding affinity or specificity. Efficacy of targeting is being enhanced through the bypassing of biological barriers.

7.12 Conclusion

There is a significant need of ultrasensitive diagnostic tool which can perceive the sub-molecular deformities, at biochemical or hereditary level. Most of the current diagnostic tools are unfit because of the nonappearance of association between the particle of interest and sensor. This is a result of almost more noteworthy size of the sensor substances used in quantifying and evaluating. In basic words, there ought to be a nearby cooperation between the particle (of interest, like seeds or other planting material) and the one which is used as a sensing. Biological systems are believably composed with functional nano-metric devices like nucleic acids, enzymes and motor proteins, which are made of brilliant assemblage of macromolecules. These mystic tools work with great precision in order to drive most insightful bio processes, like DNA replication, protein folding, temporal and spatial gene expression, movement of the cell (assisted by extracellular matrix) and cellular proliferation. It remained a huge task to detect the molecular signature markers of a specific disease because of very low concentration of observable objects. Another chief reason is the lack of sensitive tools for indicating the small amount of fungal, viral or bacterial infections. Here nanotechnology can play its role.

References

Abd El-Hai KM, El-Metwally MA, El-Baz SM (2010) Reduction of soybean root and stalk rots by growth substances under salt stress conditions. Plant Pathol J 91:149–161

Ai T, Zhang L, Gao Z, Zhu CX, Guo X (2011) Highly efficient virus resistance mediated by artificial microRNAs that target the suppressor of PVX and PVY in plants. Plant Biol 13(2):304–316

Alghuthaymi MA, Almoammar H, Rai M, Said-Galiev E, Abd-Elsalam KA (2015) Myconanoparticles: synthesis and their role in phytopathogens management. Biotechnol Biotechnol Equip 29(2):221–236

Ali I, Amin I, Briddon RW, Mansoor S (2013) Artificial microRNA-mediated resistance against the monopartite begomovirus Cotton leaf curl Burewala virus. Virol J 10(1):231

Ali S, Rodriguez-Algaba J, Thach T, Sorensen CK, Hansen JG, Lassen P, Nazari K, Hodson DP, Justesen AF, Hovmoller MS (2017) Yellow rust epidemics worldwide were caused by pathogen races from divergent genetic lineages. Front Plant Sci 8:1057

Amini B, Kamali M, Salouti M, Yaghmaei P (2017) Fluorescence bio-barcode DNA assay based on gold and magnetic nanoparticles for detection of Exotoxin A gene sequence. Biosens Bioelectron 92:679–686

Axtell MJ, Westholm JO, Lai EC (2011) Vive la différence: biogenesis and evolution of microRNAs in plants and animals. Genome Biol 12:221

Azam A, Ahmad AS, Oves M, Khan MS, Memic A (2012) Size dependent antimicrobial properties of Cuo nanoparticles against Gram-positive and negative bacterial strains. Int J Nanomedicine 7:3527–3535

Bakhori NM, Yusof NA, Abdullah AH, Hussein MZ (2013) Development of a fluorescence resonance energy transfer (FRET)-based DNA biosensor for detection of synthetic oligonucleotide of Ganoderma boninense. Biosensors 3(4):419–428

Bao YP et al (2006) Detection of protein analytes via nanoparticle-based bio bar code technology. Anal Chem 78:2055–2059

Branton D, Deamer DW, Marziali A, Bayley H, Benner SA, Butler T, Di Ventra M, Garaj S, Hibbs A, Huang X, Jovanovich SB et al (2008) The potential and challenges of nanopore sequencing. Nat Biotechnol 26(10):1146–1153

Bryaskova R, Pencheva D, Nikolov S, Kantardjiev T (2011) Synthesis and comparative study on the actimicrobial activity of hybrid materials based on silver nanoparticles (AGNps) stabilized by polyvinylpyrrolidone (PVP). J Chem Biol 4:185–191

Cai B, Huang L, Zhang H, Sun Z, Zhang Z, Zhang GJ (2015) Gold nanoparticles-decorated graphene field-effect transistor biosensor for femtomolar MicroRNA detection. Biosense Bioelectron 74:329–334

Chaudhary V, Jangra S, Yadav NR (2018) Nanotechnology based approaches for detection and delivery of microRNA in healthcare and crop protection. J Nanobiotechnol 16:40. https://doi.org/10.1186/s12951-018-0368-8

Chen C, Ridzon DA, Broomer AJ, Zhou Z, Lee DH, Nguyen JT, Barbisin M, Xu NL, Mahuvakar VR, Andersen MR, Lao KQ (2005) Real-time quantification of microRNAs by stem–loop RT–PCR. Nucleic Acids Res 33(20):179–e179

Clarke J, Wu HC, Jayasinghe L, Patel A, Reid S, Bayley H (2009) Continuous base identification for single-molecule nanopore DNA sequencing. Nat Nanotech 4(4):265

Cabiscol E, Tamarit J, Ros J (2007) Oxidative stress in bacteria and protein damage by reactive oxygen species. Int Microbiol 3:3–8

Dameron CT, Reeser RN, Mehra RK, Kortan AR, Carroll PJ, Steigerwaldm ML, Brus LE, Winge DR (1989) Biosynthesis of cadmium sulphide quantum semiconductor crystallites. Nature 338(6216):596–597

Degliangeli F, Kshirsagar P, Brunetti V, Pompa PP, Fiammengo R (2014a) Absolute and direct microRNA quantification using DNA–gold nanoparticle probes. J Am Chem Soc 136(6):2264–2267

Degliangeli F, Pompa PP, Fiammengo R (2014b) Nanotechnology based strategies for the detection and quantification of microRNA. Chemistry 20(31):9476–9492

Duan CG, Wang CH, Fang RX, Guo HS (2008) Artificial microRNAs highly accessible to targets confer efficient virus resistance in plants. J Virol 82(22):11084–11095

Edmundson MC, Capeness M, Horsfall L (2015) Exploring the potential of metallic nanoparticles within synthetic biology. N Biotechnol 31(6):572–578

Egan AN, Schlueter J, Spooner DM (2012) Applications of next-generation sequencing in plant biology. Am J Bot 99(2):175–185

Fahim M, Millar AA, Wood CC, Larkin PJ (2012) Resistance to Wheat streak mosaic virus generated by expression of an artificial polycistronic microRNA in wheat. Plant Biotechnol J 10(2):150–163

Garg J, Poudel B, Chiesa M, Gordon JB, Ma JJ, Wang JB, McKinley GH (2008) Enhanced thermal conductivity and viscosity of copper nanoparticles in ethylene glycol nanofluid. J Appl Phys 103(7):074301

Goluch ED, Nam JM, Georganopoulou DG, Chiesl TN, Shaikh KA (2006) A bio bar code assay for on-chip attomolar-sensitivity protein detection. Lab Chip 6(10):1293–1299

Gurunathan S, Han JW, Dayem AA, Eppakayala V, Kim JH (2012) Oxidative stress-mediated antibacterial activity of graphene oxide and reduced graphene oxide in *Pseudomonas aeruginosa*. Int J Nanomedicine 7:5901–5914

Guzman M, Dille J, Godet S (2009) Synthesis of silver nanoparticles by chemical reduction method and their antibacterial activity. Int J Chem Biomol Eng 2(3):104–111

Hayden CE (2015) Pint-sized DNA sequencer impresses first users. Nat News 521(7550):15

He L, Liu Y, Mustapha A, Lin M (2011) Antifungal activity of zinc oxide nanoparticles against *Botrytis cinerea* and *Penicillium expansum*. Microbiol Res 166(3):207–215

Hsueh PR (2010) New Delhi metallo-β-lactamase-1 (NDM-1): an emerging threat among Enterobacteriaceae. J Formos Med Assoc 109(10):685–687

Hu Q, Wujcik EK, Kelarakis A, Cyriac J, Gong X (2017) Carbon-based nanomaterials as novel nanosensors. J Nanomater 2017:1–2

Jain K (2003) Nanodiagnostics: application of nanotechnology (NT) in molecular diagnostics. Expert Rev Mol Diagn 3(2):153–161

Jain PK, El-Sayed IH, El-Sayed MA (2007) Au nanoparticles target cancer. Nano Today 2(1):18–29

Jayaseelan C, Rahuman AA, Kirthi AV, Marimuthu S, Santhoshkumar T (2012) Novel microbial route to synthesize ZnO nanoparticles using *Aeromonas hydrophila* and their activity against pathogenic bacterial and fungi. Spectrochim Acta A Mol Biomol Spectrosc 90:78–84

Jelly NS, Schellenbaum P, Walter B, Maillot P (2012) Transient expression of artificial microRNAs targeting Grapevine fan leaf virus and evidence for RNA silencing in grapevine somatic embryos. Transgenic Res 21(6):1319–1327

Jo YK, Kim B, Jung G (2009) Antifungal activity of silver ions and nanoparticles on phytopathogenic fungi. Plant Dis 93:1037–1043

Joseph T, Morrison M (2006) Nanotechnology in agriculture and food. Institute of Nanotechnology, London

Jurgons R, Seliger C, Hilpert A, Trahms L, Odenbach S, Alexiou C (2006) Drug loaded magnetic nanoparticles for cancer therapy. J Phys Condens Matt 18(38):S2893

Kashyap PL, Kumar S, Singh R, Kumar A et al (2013) LAMP for detection of plant pathogens. AGROBIOS Newlett XII:76–77

Khan MR (2012) Nematode, an emerging threat to global forest: assessment and management. Plant Pathol J 11:99–113

Khan MR, Jairajpuri MS (2010a) Nematode infestation in fodder crops national scenario. In: Khan MR, Jairajpuri MS (eds) Nematode infestation Part 1: Food crops. National Academy of Science, Allahabad, pp 1–16. ISBN-13:9788190554824

Khan MR, Jairajpuri MS (2010b) Nematode infestation in industrial crops–national Scenaria. In: Khan MR, Jairajpuri MS (eds) Nematode infestation. Part II: Industrial crops. National Academy of Science, Allahabad, pp 1–9. ISBN-13: 9788190554824

Khan MR, Majid S, Mohiddin FA, Khan N (2011) A new bioprocess to produce low cost powder formulations of biocontrol bacteria and fungi to control fusarial wilt and root-knot nematode of pluses. Biol Control 57:130–140

Khiyami MA, Hassan A, Yasser MA, Mousa AA, Kamel AAE (2014) Plant pathogen diagnostic techniques: forthcoming changes? Biotechnol Biotechnol Equip 28(5):775–785

Kim SW, Kim KS, Lamsal K, Kim YJ, Kim SB, Jung M (2009) An *in vitro* study of the antifungal effect of silver nanoparticles on oak wilt pathogen Raffaelea sp. J Microbiol Biotechnol 19:760–764

Kis A, Tholt G, Ivanics M, Várallyay É, Jenes B, Havelda Z (2016) Polycistronic artificial miRNA mediated resistance to W heat dwarf virus in barley is highly efficient at low temperature. Mol Plant Pathol 17(3):427–437

Knudsen BR, Jepsen ML, Ho YP (2013) Quantum dot-based nanosensors for diagnosis via enzyme activity measurement. Expert Rev Mol Diagn 13(4):367–375

Krishnaraj C, Ramachandran R, Mohan K, Kalaichelvan PT (2012) Optimization for rapid synthesis of silver nanoparticles and its effect on phytopathogenic fungi. Spectrochim Acta A Mol Biomol Spectrosc 93:95–99

Kumar S, Tao C, Chien M, Hellner B, Balijepalli A, Robertson JW, Ju J (2012) PEG-labeled nucleotides and nanopore detection for single molecule DNA sequencing by synthesis. Sci Rep 2:684

Kundu A, Saha S (2014) *In vitro* antifungal activity of noval Picolibnamide 3s against soil borne fungi and strcture activity relationship. Plant Pathol J 13:152–159

Kung YJ, Lin SS, Huang YL, Chen TC, Harish SS, Chua NH, Yeh SD (2012) Multiple artificial microRNAs targeting conserved motifs of the replicase gene confer robust transgenic resistance to negative sense single stranded RNA plant virus. Mol Plant Pathol 13(3):303–317

Kurbanoglu S, Ozkan SA (2018) Electrochemical carbon based nanosensors: a promising tool in pharmaceutical and biomedical analysis. J Pharm Biomed Anal 147(5):439–457

Latifa A, Idriss T, Hassan B, Amine SM, El Hassane B, Abdellah ABA (2011) Effects of organic acids and salts on the development of *Penicillium italicum*: the causal agent of citrus blue mold. Plant Pathol J 10(3):99–107

Lattanzio VMT, Nivarlet N, Lippolis V, Gatta SD et al (2012) Multiplex dipstick immunoassay for semi-quantitative determination of *Fusarium* mycotoxins in cereals. Anal Chim Acta 718:99–108

Lin SS, Wu HW, Elena SF, Chen KC, Niu QW, Yeh SD, Chen CC, Chua NH (2009) Molecular evolution of a viral non-coding sequence under the selective pressure of amiRNA-mediated silencing. PLoS Pathog 5(2):100–312

Liu WT (2006) Nanoparticles and their biological and environmental applications. J Biosci Bioeng 102(1):1–7

Liu Q, Chen B, Wang Q, Shi X, Xiao Z, Lin J, Fang X (2009) Carbon nanotubes as molecular transporters for walled plant cells. Nano Lett 9:1007–1010

Louder JK (2015) Nanotechnology in agriculture: interactions between nanomaterials and cotton agrochemicals. Dissertation, Texas Tech University, Texas, USA

Ma X, Tang Z, Qin J, Meng Y (2015) The use of high-throughput sequencing methods for plant microRNA research. RNA Biol 12:709–719

Mccandless L (2011) Nanotechnology offers new insights to plant pathology. College of Agricultural and Life Sciences, Cornell University, pp 17–18

McNeil SE (2005) Nanotechnology for the Biologist. J Leukoc Biol 78:585–594

Mitter N, Zhai Y, Bai AX, Chua K, Eid S, Constantin M, Mitchell R, Pappu HR (2016) Evaluation and identification of candidate genes for artificial microRNA-mediated resistance to tomato spotted wilt virus. Virus Res 211:51–158

Mornet S, Vasseur S, Grasset F, Duguet E (2004) Magnetic nanoparticle design for medical diagnosis and therapy. J Mater Chem 14:2161–2217

Nair R, Kumar SD (2013) Chapter 10. Plant diseases—control and remedy through nanotechnology. In: Tuteja N, Gill SS (eds) Crop improvement under adverse conditions. Springer Science+Business Media, New York, NY. https://doi.org/10.1007/978-1-4614-4633-0_10

Nam JM, Thaxton CS, Mirkin CA (2003) Nanoparticle-based bio-bar codes for the ultrasensitive detection of proteins. Science 301:1884–1886

Nam JM, Stoeva SI, Mirkin CA (2004) Bio-bar-code-based DNA detection with PCR-like sensitivity. J Am Chem Soc 126(19):5932–5933

Nezhad AS (2014) Future of portable devices for plant pathogen diagnosis. Lab Chip 14:2887–2904

Niu QW, Lin SS, Reyes JL, Chen KC, Wu HW, Yeh SD, Chua NH (2006) Expression of artificial microRNAs in transgenic *Arabidopsis thaliana* confers virus resistance. Nat Biotech 24(11):1420

Oh SD, Lee S, Choi SH, Lee IS, Lee YM, Chun JH, Park HJ (2006) Synthesis of Ag and Ag-SiO_2 nanoparticles by gamma irradiation and their antibacterial and antifungal efficiency against *Salmonella enterica*, *Serovar typhimurium* and *Botritis cinerea*. Colloid Surf A Physiochem Eng Asp 275:228–233

Pan L, Qiu H, Dou C, Li Y, Pu L, Xu J, Shi Y (2010) Conducting polymer nanostructures: template synthesis and applications in energy storage. Int J Mol Sci 11:2636–2657

Panacek A, Kolář M, Večeřová R, Prucek R, Soukupova J, Kryštof V, Kvítek L (2009) Antifungal activity of silver nanoparticles against *Candida* spp. Biomaterials 30(31):6333–6340

Park HJ, Kim SH, Kim HJ, Choi SH (2006) A new composition of nanosized silica-silver for control of various plant diseases. Plant Pathol J 22:295–302

Pimentel D (2009) Invasive plants: their role in species extinctions and economic losses to agriculture in the USA. In: Management of invasive weeds, invading nature – Springer Series in invasion ecology. Springer, Dordrecht, pp 1–7

Răcuciu M, Creanga DE (2009) Biocompatible magnetic fluid nanoparticles internalized in vegetal tissue. Rom J Phys 54:115–124

Răcuciu M, Miclaus S (2007) Low-level 900 MHz electromagnetic field influence on vegetal tissue. Rom J Biophys 17(3):149–156

Rad F, Mohsenifar A, Tabatabaei M, Safarnejad MR et al (2012) Detection of Candidatus *Phytoplasma aurantifolia* with a quantum dots FRET-based biosensor. J Plant Pathol 94(3):525–534

Rosi NL, Mirkin CA (2005) Nanostructures in biodiagnostics. Chem Rev 105(4):1547–1562

Sadanandom A, Napier RM (2010) Biosensors in plants. Curr Opin Plant Biol 13(6):736–743

Safarpour H, Safarnejad MR, Tabatabaei M, Mohsenifar A et al (2012) Development of a quantum dots FRET-based biosensor for efficient detection of *Polymyxa betae*. Can J Plant Pathol 34(4):507–515

Sahayaraj K (2012) Bionanomaterials: synthesis and applications. In: Proceedings of the first national seminar on new materials research and nanotechnology (NSNMRN 2012), Government Arts College, Ooty, Tamil Nadu, 12–14 September 2012

Sahayaraj K, Roobadevi M, Rajesh S, Azizi S (2015) *Vernonia cinerea* (L.) Less. silver nanocomposite and its antibacterial activity against a cotton pathogen. Res Chem Interm 41:5495–5507

Shu Y, Shu D, Haque F, Guo P (2013) Fabrication of pRNA nanoparticles to deliver therapeutic RNAs and bioactive compounds into tumor cells. Nat Protoc 8:1635–1659

Singh M, Singh S, Prasada S, Gambhir IS (2008) Nanotechnology in medicine and antibacterial effect of silver nanoparticles. Digest J Nanomater Biostruct 3:115–122

Siva C, Kumar MS (2015) Pesticidal activity of eco-friendly synthesized silver nanoparticles using *Aristolochia indica* extract against *Helicoverpa armigera* Hubner (Lepidoptera: Noctuidae). Int J Adv Sci Tech Res 2:197–226

Solgi M, Kafi M, Taghavi TS, Naderi R (2009) Essential oils and silver nanoparticles (SNP) as novel agents to extend vase-life of gerbera (*Gerbera jamesonii* cv. 'Dune') flowers. Postharvest Biol Technol 53:155–158

Stanisavljevic M, Son K, Vaculovicova M, Kizeka R, Adama V (2015) Quantum dots-fluorescence resonance energy transfer-based nanosensors and their application. Biosens Bioelectron 74(15):562–574

Sun L, Lin C, Du J, Song Y, Jiang M, Liu H, Zhou S, Wen F, Zhu C (2016) Dimeric artificial microRNAs mediate high resistance to RSV and RBSDV in transgenic rice plants. Plant Cell Tiss Org Cult 126(1):127–139

Talibi I, Askarne L, Boubaker H, Boudyach EH, Aoumar AAB (2011) *In vitro* and *in vivo* antifungal activities of organic and inorganic salts against citrus sour rot agent *Geotrichum candidum*. Plant Pathol J 10:138–145

Thind TS (2012) Phytopathogenic Procaryotes and plant diseases. Plant Dis Res 27(1):125–125

Van V, Choudhury TNR, Mukherje SK (2013) Transgenic tomato plants expressing artificial microRNAs for silencing the pre-coat and coat proteins of a begomovirus, Tomato leaf curl New Delhi virus, show tolerance to virus infection. Virus Res 172(1–2):35–45

Wagaba H, Patil BL, Mukasa S, Alicai T, Fauquet CM, Taylor NJ (2016) Artificial microRNA-derived resistance to Cassava brown streak disease. J Virol Methods 231:38–43

Wang C, Zhao M, Li J, Yu J, Sun S, Ge S, Huang Y (2017a) Silver nanoparticles/graphene oxide decorated carbon fiber synergistic reinforcement in epoxy-based composites. Polymer 131:263–271

Wang J, Qiufang Y, Chengmin S, Chunde J, Qingfeng S (2017b) One-step preparation of graphene oxide/cellulose nanofibril hybrid aerogel for adsorptive removal of four kinds of antibiotics. J Nanomater 2017:5150613. https://doi.org/10.1155/2017/5150613. 10 pages

Wang L, Hu C, Shao L (2017c) The antimicrobial activity of nanoparticles: present situation and prospects for the future. Int J Nanomed 12:1227–1249

Wiesner MR, Lowry GV, Alvarez P, Dionysiou D, Biswas P (2006) Assesing the risk of manufactured nanomaterials. Environ Sci Technol 15:4336–4345

Xiong T, Yuan X, Wang H, Leng L, Li H, Wu Z, Zeng G (2018) Implication of graphene oxide in Cd-contaminated soil: a case study of bacterial communities. J Environ Manage 205:99–106

Xu J (2016) Fungal DNA barcoding. Genome 59(11):913–932

Younes H, Rahman Md M, Ghaferi AA, Saadat I (2017) Effect of saline solution on the electrical response of single wall carbon nanotubes-epoxy nanocomposites. J Nanomater 2017:6843403. https://doi.org/10.1155/2017/6843403. 8 pages

Yadav A, Kon K, Kratosova G, Duran N, Ingle AP, Rai M. (2015) Fungal as an efficent mycosin for the synthesis of metal nanoparticles: progress and key aspects of resrach. Biotech lett. 37(11):2099–2120

Zhang X, Li H, Zhang J, Zhang C, Gong P, Ziaf K, Xiao F, Ye Z (2011) Expression of artificial microRNAs in tomato confers efficient and stable virus resistance in a cell-autonomous manner. Transgenic Res 20(3):569–581

Zhao MX, Zeng EZ (2015) Application of functional quantum dot nanoparticles as fluorescence probes in cell labeling and tumor diagnostic imaging. Nanoscale Res Lett 10:171

Zhang L, Jiang Y, Ding Y, Daskalakis N, Jeuken L, Povey M, O'Neill AJ, York DW (2010) Mechanism investigation into antibacterial behaviour of suspension of ZnO nanoparticles against E.coli. J Nanoparticle Res 12(5):1625–1636

Chapter 8
Nanofertilizers

Beenish Zia Butt and Iqra Naseer

8.1 Introduction

Agriculture plays a crucial role in the world economy, and is known as the backbone of several developing countries. The extensive and relentless application of agro-chemicals in the agricultural cropping systems to secure greater yield has led to many health risks and environmental concerns. Regardless of all the associated environmental problems, chemical fertilizers are being frequently used to achieve more food with a better quality, in order to feed the world's ever growing population (Ghaly 2009). The identification and use of some new substituent compounds instead of chemical fertilizers is an urgent demand of the present time, though the accomplishment of sustainable agriculture through the exploitation and development of nano-technological advancements seems to be a capable approach (Pijls et al. 2009). The development and use of nano based materials including nanofertilizers and nanopesticides is one of the promising solutions of all agricultural problems. Extremely small size, large surface area to volume ratio, surface specificities and many other unique characteristics make nano materials as magical tools and their application in agriculture can be carried out to achieve a balanced and highly nutritive food (Joseph and Morrison 2006).

B. Z. Butt (✉)
Monash University, Subang Jaya, Malaysia

I. Naseer
Department of Botany, Lahore College for Women University, Lahore, Pakistan

S. Javad (ed.), *Nanoagronomy*, https://doi.org/10.1007/978-3-030-41275-3_8

8.1.1 Chemical Fertilizers in Agricultural Cropping Systems

Chemical fertilizers are inorganic materials or compounds of mainly synthetic origin and have a specific chemical composition. In agricultural cropping system these chemical fertilizers are mainly used as a nutrient supply like potassium, nitrogen and phosphorus which are deficient in soil. But the key concern in their application is their loss from soil through leaching, water runoff and volatilization of some compounds. The release of chemical nutrients from conventional chemical fertilizers is also a main cause of degradation of environment, a considerable amount of nutrients is lost to the environment and hence become unavailable for plants (Ombódi and Saigusa 2000; Trenkel 1997).

Conventional chemical fertilizers are extensively used in agricultural systems, applied usually through fumigation. Though, the actual efficacy of these fertilizers depends on the actual concentration of the chemical nutrient that is absorbed by the plant. In actual only a small percentage of the applied fertilizers are absorbed by the plant roots and most of the proportion is lost to the environment. Therefore, their mode of application is also a main factor which should be considered to reduce the environmental pollution. The frequent use of agrochemicals in agricultural practices has also been proved to be a major cause of soil nutrient deficiency and imbalance, besides this, it adversely affects the plants and microbial community of the soil (Ombódi and Saigusa 2000).

Several devastating effects of chemical fertilizers which are degrading the environmental quality have also been reported. Thus to achieve the maximum production there is a great need to optimize the over exploitation of fertilizers which in turn will also reduce the pollution (Tilman et al. 2002).

Hence the use and development of other most efficient alternatives like nanofertilizers which would have an excellent impact on plant and soil properties is a great need of the time (Miransari 2011).

8.1.2 Nano Agriculture

Nanotechnology has provided the feasibility of exploring nanoscale materials as fertilizer carrier or fertilizer by themselves which can be used for the fabrication of smart fertilizers. These nano structured fertilizers used to enhance the nutrient use efficiency and reduce the cost of environmental pollution (Chinnamuthu and Boopathi 2009). The use of nano based fertilizers seems to be a significant alternative to meet the nutrient requirements of the plants and agricultural crops. The application of nano coated fertilizers provides a slow release of nutrients which are readily absorbed by the plants. In last few years, the development and exploitation of nanofertilizers has gained much importance over the chemical fertilizers which are heavily degrading the environment by causing the land pollution (Wu and Liu 2008).

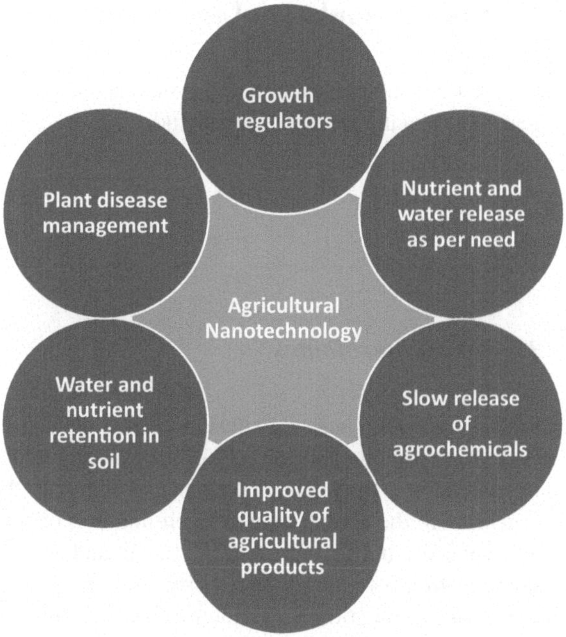

Fig. 8.1 Different aspects of nanotechnology in agriculture

Table 8.1 Some advantages of nanofertilizers over the use of conventional fertilizers (Miransari 2011)

Sr. No	Properties	Nano-fertilizers	Conventional fertilizers
1	Solubility	High solubility	Low solubility
2	Adsorption capacity	Lesser	higher
3	Bioavailability	High	Less
4	Nutrient uptake efficiency	High nutrient uptake	Low nutrient uptake
5	Release of nutrients	Slow and controlled release	Rapid release
6	Loss rate	Less	High

In general, nano materials that are mainly used in crop nutrient delivery system are referred as nano fertilizers. Nano material coated nutrient encapsulations or emulsions of nano materials are examples of nanofertilizers (DeRosa et al. 2010). Nano coated nutrient particles are more efficient in holding nutrients more strongly due to their unique surface characteristics which aid in site specific and slow nutrient release than the usual material surfaces that are used in chemical fertilizers synthesis (Brady and Well 1999) (Fig. 8.1).

One of the most prime aspects of nanotechnology in agriculture is the release and supply of macro- and micronutrients through chemical fertilizers (Table 8.1). These characteristics features of nano materials are because of their great surface area and active sites, stability, extremely smaller dimensions and least toxic effects are attributed to their enormous exploitation in agriculture (Sasson et al. 2007).

8.2 Nano Formulation Used in Agriculture System

These are the nano materials-based products used to attain a sustainable agriculture system. These nano formulations includes

- Nanoherbicides
- Nanofungicides
- Nanofertilizers

8.2.1 *Nanoherbicides*

Herbicides are the chemical substances that are used to eliminate and reduce the growth of unwanted plants (weeds). The presence of these superfluous weeds in the main cultivated greatly affects the nutrient supply to the main crop plant and results in a decline in the actual yield. The elimination of these weeds through herbicides can be achieved but still to avoid the health and environmental risks a least or optimal application is required. Nanoherbicides are the alternative forms of herbicides, prepared in combination with nano materials, polymeric core shell nanoparticles are the encapsulated herbicides (Kumar et al. 2015). Their efficient mode of action significantly controls the growth of parasitic weeds. Encapsulations of target specific herbicide in nanoparticles are used against specific weeds, these acts on particular root receptors and kill the target weed (Jampílek and Kralova 2015).

8.2.2 *Nanofungicides*

These are the nano formulation used to manage the plant fungal pathogens in agriculture cropping system. Essential oil encapsulations in nanoparticles have potent antifungal activity. Chitosan nanoparticles encapsulated with essential oil of *Zataria multiflora* were studied for their antifungal activity against Botrytis, the slow release of oil from these encapsulations showed a great reduction in disease occurrence (Mohammadi et al. 2015). Similarly, thyme oil containing polyethylene terephthalate punnets encapsulated with chitosan nanoparticles demonstrated a notable decline in brown rot of peach (Jampílek and Kralova 2015) (Table 8.2).

8.2.3 *Nanofertilizers*

These are the altered forms of conventional chemical fertilizers, fabricated through various *physicochemical* and biological methods with the help of nanotechnological interventions. Nanofertilizers have distinctive novel characteristics than the bulk

Table 8.2 Applications of nanotechnology (Manjunatha et al. 2016)

Sr. No	Delivery	Application	Nano materials
1	Nano fertilizers	RNA (double stranded)	Histosan
		Controlled delivery of NPK	Chitosan, sulfur nanocoatings
2	Biopesticides	Encapsulated essential oils (*Artemisia arborescens*)	Solid lipid
		Insecticide nanosilica	Nanosilica

materials and also have enormous advantages over the traditional fertilizers and used to improvise the crop production and soil properties in general (Brunner et al. 2006).

8.3 Basic Forms of Nanofertilizers

These are the commonly occurring form of nanofertilizers which are greatly exploited in the agriculture systems. These are as follows:

- Zeolites
- Nano-composites
- Super absorbent fertilizers
- Carbon nanotubes

8.3.1 Zeolites

These are microscopic, crystalline hydrated forms of aluminosilicate minerals that are frequently used as adsorbents. There are different types of mineral zeolites (clinoptilolite, stilbite, chabazite, analcime, natrolite) depending upon the ion exchange capacity and ability to retain or lose water.

8.3.1.1 Zeolites as Efficient Nutrient Carriers

Zeolites being an illustrious carrier of nutrients, these also act as an additive and regulator of several nutrient mineral fertilizers (Bagdasarov et al. 2004). The application of zeolites in combination with potassium and phosphorus compounds in agriculture and horticulture is a good way to control the slow release of these nutrients (Dwairi 1998). Efficiency of nutrient use can be improved by loading zeolites to the nitrogen and potassium containing fertilizers (Ming and Allen 2001).

8.3.1.2 Beneficial Impacts of Zeolites Containing Fertilizers on Plants

Zeolites containing fertilizers are good source of nutrients which hold and release nutrients slowly (nitrogen, calcium, potassium, magnesium etc.) in the rhizosphere which are easily extracted by the plants (Flanigen and Mumpton 1981). As a result, notable increase in efficiency of potassium and nitrogen use occurs which leads to a remarkable increase in crops yield.

Application of these slow release fertilizers in the initial stages of seedlings has profound effects on the agricultural crops development and yield (Arthanareeswaran et al. 2008). Growth promoting effects of calcium containing zeolites were also observed on tomato and rice crop production, as a decline in sodium ions concentration in plants shoot was recorded (Song and Fujiyama 1996). Effects of zeolites added chemical fertilizers was observed on the yield and production of grapes, peach and tomato (Burriesci et al. 1984; Valente et al. 1982). The application of chemical fertilizers in combination with zeolites was assessed to investigate the nutrient use efficiency by maize. The addition of these slow release fertilizers significantly improved the potassium, nitrogen and phosphorous uptake by the plant parts. Results established that zeolites have several advantages over the traditional chemical fertilizers (Ahmed et al. 2010).

Application of zeolites in improving the photosynthesis and plant biomass was also reported in barley and maize (Krutilina et al. 2000). The use of zeolites due to their excellent cation exchange capacity has also been investigated for the production of most important crops including vegetables, cereals, fruits (Butorac et al. 2002). Zeolites in combination with fertilizers can be used to increase the nutrient use efficiency by the crop with ultimately effect on the plant growth and results in stunning production (Polat et al. 2004). Similarly the use of zeolites has been substantiated to improvise the Nitrogen uptake by plants, its use efficiency and biomass production by reducing the loss of ammonia to the atmosphere (He et al. 2002).

8.3.1.3 Zeolites as Soil Conditioning Agents

Zeolites act as soil conditioner which improves the soil structure and other properties in general. The amendments of zeolite containing fertilizers has also been studied and reported for their remarkable moisture holding, ion exchange capacity and growth promoting potential. Saline soils have a high percentage of Na ions which interferes with the overall nutrient content of the soil. Addition of zeolites was found to be a good source to balance the sodium ratio by supplying calcium ions in saline soils. The release of calcium checks the sodium accumulation in the rhizosphere which has harmful effects of the plant development (Ayan et al. 2005).

8.3.1.4 Nano Porous Zeolite

Nano porous zeolites are an excellent source of slow release nutrient fertilizers, which reduce the mineral nutrient loss and release the required nutrients as per need of plants. Their large surface area and highly active sites make them suitable for nutrient substitutions which are replaced by other ions through cation exchange process (Naderi and Danesh-Shahraki 2013).

Being an essential element nitrogen is a vital nutrient for plants, but the greater solubility of nitrogen containing fertilizers can have serious damaging effects on plants, though a slow release and significant increase in nitrogen uptake from urea can be achieved by using urea in combination with zeolites (Manik and Subramanian 2014).

8.3.2 Nano-composites

These are multiphase matrixes of silicates (montmorillonite clay and some organic polymers), which are incorporated with nanoparticles to improve the desired property of the material. It is anticipated that these are the promising tools that can be used to develop nano materials with unique characteristics. The application of ammonium loaded clinoptilolite in combination with phosphorite as a good carrier and facilitate the slow release of nitrogen and phosphorus (Barbarick et al. 1990). The combination of phosphorite, potassium and ammonium loaded clinoptilolite has been reported for the slow release of Ca, P, K, N (Allen 1991).

8.3.3 Super Absorbent Fertilizer (SAF)

These are the copolymers of fertilizers (nitrogen, potassium, phosphorus) and super absorbent polymer, these nanofertilizers are linked through hydrogen bonds. The water holding potential of these super absorbent fertilizers largely depends on the hydrophilic groups, present in their molecular structure. Slow release of nitrogen and the moisture holding capacity of soil was also studied using super absorbent nitrogen fertilizer (SANF). The study substantiated that the products having cross linkages are excellent slow releasers of nutrients and have a high-water retention capacity which in turn enhances the water and fertilizer use efficiency by the crop. In arid regions SSNF have remarkable application in agricultural practices (Liu et al. 2006).

The effects of nanofertilizers (incorporated and coated with nanoparticles) in comparison with chemical fertilizer were observed on the yield and nutrient contents of wheat. Results indicated an increase in the overall yield and protein content of wheat. It was confirmed that the use of nanofertilizers (slow release fertilizers) can be made to improvise the nutritional value and production of wheat (Xiao et al. 2008).

8.3.4 *Carbon Nano Tubes (CNTs)*

These are the nano structured, allotropic forms of carbon, which possess several unique properties that facilitate their extensive use in agricultural industry. The use of carbon nano tubes as slow release fertilizers have excellent effects on seeds germination, results confirmed up to twofold increase in germination rate and seeding biomass was recorded in experimental plants than the control plants. Furthermore, analysis of the seed revealed the penetration of carbon nano tubes inside the seed which in turn have ameliorating effects on germination and growth (Khodakovskaya et al. 2009).

The effects of CNTs on water absorption and retention capacity of seeds have also been studied. Water content in the CNTs incubated seeds was calculated as 57.6% though it was 38.9% for seed in control. Results elucidated the formation of new pores (water gating channels) in the seed coat due to the penetration of CNTs, caused a higher absorption (Srinivasan and Saraswathi 2010). Nanotechnology is considered as an essential and promising area of science that can make a great progress in our agricultural products. Most of the studies that have been done so far in recent years inclusively encouraging the exploitation of slow release fertilizers and zeolites in the agricultural fields to improve the production (Dutta and Sugunan 2004).

8.4 Classification of Nanofertilizers on the Basis of Mineral Nutrients

Mineral nutrients play an essential role in plant growth and metabolism. There are two main types of nano nutrient fertilizers.

- Macro nanofertilizers
- Micro nanofertilizers

8.4.1 *Macro Nanofertilizers*

Macro nutrients are the mineral elements which are required by the plants in large amounts. These include nitrogen (N), phosphorus (P) and potassium (K). Macronutrients are further categorized into primary and secondary nutrients.

8.4.1.1 Primary Nutrients Coated Nanofertilizers

Nitrogen Nanofertilizer

Nitrogen is a vital chemical element of several structures; proteins are the main structural units of plant cells made up of amino acids which possess nitrogen as an important constituent. Being an essential component of the chlorophyll, DNA and ATP, it takes part in many metabolic and regulatory pathways. Three basic forms of nitrogen are readily accessible to plants, this includes nitrate and ammonium ions and organic nitrogenous substances, though most of the atmospheric nitrogen is unavailable and plants cannot utilize this nitrogen (Preetha and Balakrishnan 2017).

Nitrogen Leaching

Leaching is the main difficulty that is encountered by the farmers during the application of fertilizers, to overcome this loss few modified forms of fertilizers like neem and sulfur coated urea and polyolefin resin coated urea are used which lowers the nitrogen release. Though the use of such kind of fertilizers is a costly approach, substitution of different cation exchangers in chemical fertilizers is a significant way to control the NH^{4+} leaching.

Nitrogen Zeolites

Use of zeolites is a very effectual alternative that retains and releases the required nutrients hence it boosts the crop productivity by providing all the nutrients in appropriate amounts. Clinoptilolite zeolite (porous zeolite), has an excellent affinity for ammonium ions with a notable cation exchange capacity (300 cmol kg^{-1}) (Gupta et al. 1997). It has been applied to control the release of ammonia and protects the plants from the toxic effects of ammonium ions (Amon et al. 1997).

Soil Amendments of Zeolites

The moisture and nutrients holding capacity of sandy soils is quite low as these are coarse structured soils therefore prone to leaching. The amendments of clinoptilolite zeolite (CZ) in sandy soils have been reported to reduce the loss of nitrate and ammonium ions during leaching and also to improve the water holding capacity as zeolites offers a high surface area to volume ratio (Huang and Petrovic 1994). Urea is one of the most widely used fertilizers that have been an economical source of nitrogen, but in agricultural practices it rapidly transforms into ammonia that is readily released in the atmosphere. It is a major limiting factor which reduces the use of urea as nitrogen source. The exploitation of CZ due to its high cation exchange capacity (CEC) allows the slow and steady release of ammonia (Kithome et al. 1998).

The use of CZ coated ammonium sulfate was also reported to reduce the loss of nitrogen and to improve the nitrogen use efficiency by crop and significant results were recorded in comparison with ammonium sulfate. The application of clinoptilolite not only increases the nitrogen use efficiency by the crops though it also limits the nitrification and hence lowers the nitrate loss from soil (Perrin et al. 1998). The application of urea along with sago waste water and zeolites has been reported as an advantageous approach as it promotes the availability of nitrate and ammonium ions for plants rather ammonia (Latifah et al. 2011). The nutrient release phenomenon from nano-fertilizers were observed by several scientists, results showed that these nanofertilizers release the nutrients for an extended period of 50 days that is threefold than the usual chemical fertilizers. This indicated the use of zeolite as an appropriate approach to increase the nitrogen availability and efficiency (Sharmila 2010).

Phosphorus Nanofertilizer

Phosphorus is a vital mineral component. It is one of the other indispensables for all living beings. In plants it is one of the key components that regulate the several biochemical processes and being a structural component of many compounds including ATP and ADP, it plays crucial role in energy transfer. Phosphorus is also a main structural part of enzymes, phospholipids, nucleic acids and sugar phosphates. A sufficient quantity of phosphorus is required at premature phase for the development of reproductive structures in plants. There are several growth parameters which are directly coupled with the availability of phosphorus to crop, these include improvement in seed yield, increase in root shoot length and vigor, enhanced flowering, disease resistance and crop quality and yield (Preetha and Balakrishnan 2017).

Phosphorus Zeolites

The impact of nutrient cation coated zeolites on the solubility of phosphate rock was studied and it was reported that crop only uses 20% of the applied concentration, the other 80% becomes a part of the phosphorus pool in the soil and its gradual release becomes available to the plant over the time (Malhi et al. 2002). Recently a research was conducted on zeoponic, to assess the release of nutrient to plant in artificial nutrient growth system, zeoponic which releases the nutrients like N, P, K as per plant requirement by the ion exchange and dissolution of apatite. These artificial systems improve the nutrient retention capacity, lower the mineral nutrient loss and also reduce the fertilizers dependence by maintaining a balance of renewable supply to the plants (Preetha and Balakrishnan 2017). Phosphate release scheme from zeolites and nano clays was also reported and it has been observed that nanofertilizers release the phosphate slowly up to 1200 h while the conventional fertilizers release all the nutrients within 300 h and hence it is suggested that the use of surface modified zeolites is a promising way to improve the potassium use efficiency (Subramanian and Rahale 2009).

Potassium Nanofertilizer

Potassium as Vital Nutrient

It is an important macronutrient, required to plants in adequate amount to carry out numerous physiological and biochemical processes which are essential for plant survival. It has a basic role in protein synthesis, photosynthesis, maintain the ionic equilibrium, translocation of carbohydrates, regulate water use efficiency and also act as a catalyst in many enzymatic processes. It has been reported that it catalyzes the activation of 60 enzymes associated with the plant development and growth. Potassium deficiency causes decrease in root shoot development and reductions in seed and fruit production, the plants are more susceptible to pest and insect attach. Appropriate concentration of potassium is seemed to be an important factor that improves the color, shape, size, taste and shelf life of fruits (Preetha and Balakrishnan 2017).

Potassium Zeolites as Effective Fertilizers

Natural zeolites are known to have a capacity to absorb potassium rather other cations from fertilizers thus decrease its loss from the soil. The application of chemical commercial fertilizers in combination with zeolite showed the great amount of potassium is available in the soil (Mazur et al. 1986). From all the mineral elements potassium is one of the known nutrients that have a maximum cation exchange capacity (216 cmol kg^{-1}), therefore it is readily released and dissolved in the soil solution (Treacy and Higgins 2007). Ion exchange capacity of nano-zeolites with specific cations is mainly responsible for the gradual release of potassium from nanofertilizers. Hence, the ample amount of mineral nutrients can be provided to plants by growing these into the zeolites enriched medium (Zhou and Huang 2007).

Slow and steady release of potassium from potassium zeolites (nanofertilizer) was investigated and the effect of potassium zeolites on the different growth parameters of black pepper was also studied (Li et al. 2010). Similarly, the gradual release of potassium from nanofertilizers was also studied with respect to its effects on plant. Potassium fixation and equilibrium are mainly accountable for the accessibility/availability of potassium in the soil, though further advancements and research, nanofertilizers may be helpful in improving the crop growth and yield (Subramanian and Rahale 2009).

8.4.1.2 Secondary Nutrients Coated Nanofertilizer

Micronutrients are the secondary mineral nutrients that are required in small amounts for balanced crop nutrition. These include calcium, magnesium, sulfur. Calcium and magnesium both are coupled with the clay contents in the soil and exhibit a same behavior like potassium in the soil. The slow release of both of these nutrients has been reported by zeolites, and it is proposed that the calcium and

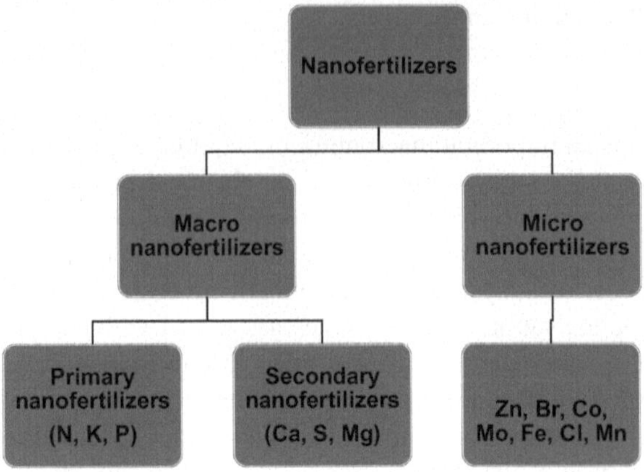

Fig. 8.2 Types of nanofertilizers

magnesium contents can be improved in the soil by the use of zeolites as fertilizers (Supapron et al. 2002).

Surface Modified Zeolite

The use of surface modified zeolite (SMZ) as stabilizer to slow down the sulfate leaching has been investigated. The experimental results showed the minimum loss of sulfate. Hence it is substantiated that the application of SMZ as additives can significantly decrease the sulfate leaching from the soil and it can be used to achieve the slow release of sulfate (Li and Zhang 2010) (Fig. 8.2).

8.4.2 *Micro Nanofertilizers*

Micronutrients are the chemical substances which are required in trace amounts for the optimal plant growth and development. These micronutrients include manganese (Mn), boron (B), molybdenum (Mo), iron (Fe), zinc (Zn), chloride (Cl) and copper (Cu). Their small amounts are as important as macronutrients for the proper plant growth and regulation.

8.4.2.1 Micronutrients Deficiency

In many countries of Asia, deficiency of these micronutrients is attributed to several abiotic factors and it is the key problem in agriculture farming. Micronutrient deficiency can have serious damaging effects on plants morphology as well as physiology which in turn causes a reduction in crop production and yield (Malakouti 2008).

8.4.2.2 Natural Zeolites

The slow release of iron and zinc was investigated from five zeolites and bentonite minerals. The results showed that natural zeolites (bentonite and chabazite minerals) are in particular more effective for nutrient sorption (Sheta et al. 2003). Foliar application of zinc and copper zeolites was reported on wheat plant. The results suggested an overall increase in protein content of the wheat plant with zinc zeolites than the plants treated with copper zeolites (Schmidt and Szakal 2007).

8.4.2.3 Zeolites as Soil Conditioning and Plant Growth Promoting Agents

The use of zeolite has been reported as soil conditioning agents, improves the soil quality by maintaining the nutrient equilibrium and also helps in the micronutrient especially cations uptake by the plant roots. An increase in the uptake of Mn and Cu was observed in zeolite/P-rock applied Sudan grass. The effect of zinc oxide nanoparticles has also been studied on the growth of rye grass, an increase in the germination and growth rate was observed with the application of ZnO nanoparticles. The potential effects of carbon nanotubes on plant cells (pumpkin crop) were also observed. The study revealed that carbon nanotubes can be used as an excellent source of nutrient supply to the plant cells (González-Melendi et al. 2007). It has been stated that ZnO NPs have a great potential and ameliorate the induction of indol acetic acid in root cells, and improve the overall plant development and yield (Pandey et al. 2010).

Boron is another micronutrient crucial for plant metabolism, its sorption phenomenon by other mineral nutrient and soil has been studied, though its accumulation in excess amounts in plant cells may have damaging effects on plant cells (Kavak 2009; Köse and Öztürk 2008; Cengeloglu et al. 2007). Molybdenum is one of the important nutrient being a part of few vital enzymes (nitrate reductase and nitrogenase) (Rana and Viswanathan 1998).

8.5 Biofertilizers and Nanobiofertilizers

Biofertilizers are the other major substances which contain living microorganisms (or latent cells of microorganisms) which are frequently used in agriculture. These increase the nutrient uptake efficiency by improving the interactions among the micro flora in the rhizosphere. These microbes include *Pseudomonas* spp., *Azospirillum*, *Bacillus* spp., *Rhizobium*, *Azotobacter*, blue green algae, and some fungal species (Wu et al. 2005). These microbes bring the conversion of organic material into much smaller simpler compounds which are a vital source of mineral nutrients and readily utilized by the plants and also maintain the soil fertility. The effectiveness of these substances depends on their handling during preparation. Application methods are also important factor to get fruitful results (Jha and Prasad

2006). Though, there are some shortcomings in the use of biofertilizers, these include short shelf life, temperature-sensitivity, and degradation during storage.

8.5.1 Nanomaterials Coated Biofertilizers

The use and preparation of nanoparticles coated biofertilizers which have a capacity to resist desiccation, has also been achieved. Emulsions like water in oil are considered as simple and achievable approach to preserve these formulations (Vandergheynst et al. 2006). The oil reduces the water loss by making a covering of water around the microorganisms and it is suitable for those microbes which are not able to resist desiccation. These emulsions have a good impact on the microbial cell viability and release. Though during preservation sedimentation may occur that is the main issue in their storage. The addition of hydrophobic silica nanoparticles in water in oil emulsions may cause a reduction in sedimentation and also increase the cell viability (Vandergheynst et al. 2007). Due to limited availability of land and water resources and development of horticultural crops of Fabaceae, use of silver and gold nanoparticles as a growth promoting materials could be effective (Dikshit et al. 2013).

8.5.2 Nanofertilizers as Growth Promoting Agents

Nano-formulation have also been studied and investigated for their growth promoting effects on many crop plants. Nanoparticles in combination with bio-fertilizers (*Bacillus subtilis*, *Paenibacilluselgii* and *Pseudomonas fluorescens*) have been reported for their growth promoting effects on the *in vitro* grown plants. These nano based bio-formulations are much useful and cost effective than the chemical fertilizers because only a small quantity of these formulations is enough for a large area application. Rhizosphere bacteria that are frequently found in the plant root zone have growth promoting effects on plants by establishing beneficial interactions among the microbial community. Dikshit et al. (2013) reported the effect of plant growth promoting rhizobacteria. Gold nanoparticles were ineffective against the *Pseudomonas putida*; however, a considerable rise was recorded in the case of *B. subtilis*, *P. fluorescens* and *P. elgii*. It was established that gold nano formulations can be applied as nanobiofertilizers (Shukla et al. 2015).

8.6 Nanofertilizers and Their Interventions in Agricultural Cropping System

8.6.1 Nano-fertilizers for Balanced Crop Nutrition

Fertilizers are responsible for enhanced food grain production from last few decades that coincide with the exponential increase in fertilizer consumptions. It has been clearly demonstrated that fertilizer contributes to the tune of 35–40% of the productivity of any crops. This results in imbalanced fertilization and occurrence of nitrate pollution in ground waters. The usage efficiencies of macronutrient fertilizers like N (30–35%), P (18–20%) and K (35–40%) remained persistent from the past few decades. Consequently, added fertilizers remained in soil or enter into aquatic system causing eutrophication (Tarafdar 2015).

The solution of various issues such as nutrient deficiencies, unavailability of organic matter and excessive fertilization is to introduce nano based fertilizers with different functions. There is a vast scope for the formation of nanofertilizers as this technology is new and revealing. Nano particles as fertilizers can be applied as foliar spray for enhanced production. It was found out that by foliar application of 640 mg ha^{-1} of nanophosphorous to cluster bean and pearl millet under arid environment yield was increased (Raliya 2012; Tarafdar 2012). Recently researchers are developing nano-composite which would supply all required nutrients in appropriate amount with the help of smart delivery system. Nanotechnology can provide balanced fertilization. When nitrogen is applied in the form of conventional fertilizers 50–70% of nitrogen is lost which causes lesser nitrogen usage efficacy of plants. Nanofertilizers are capable to positively affect the economy, energy and environment by decreasing nitrogen loss due to leaching, secretions and long term incorporation of soil microorganisms (Derosa et al. 2010). Nanoscale porous parts of plant have the capability to reduce nitrogen loss with the new nutrient delivery systems. In the future agronomy, nanofertilizers will be triggered according to the environmental conditions or timely released. Also, nanoparticles encapsulation will enhance the nutrient uptake. Similarly, controlled and slow release of fertilizers will also enhance efficiency of nutrient uptake. Coating and cementing of nanofertilizers with natural materials is more economical for crop production than to synthetic coated fertilizers. Nanofertilizers with their controlled and slow release also reduces toxicity of soil, related to higher consumption of fertilizers (Tarafdar 2015).

8.6.2 Controlled Release of Nano-fertilizers

Recently, slow and controlled release fertilizers are being developed for plants. These fertilizers in nano form ought to be efficiently more available to plants due to nano scale plant pores. Some unfavorable effects such as soil and water pollution might be caused by indiscriminate use of chemical fertilizers. Nano agrochemicals must contain all necessary properties like targeted delivery, stability, efficacy, less eco toxicity and time controlled release hence reducing repetition. With slow-release fertilizers, plants are able to take up maximum number of nutrients without leaching as compared to the soluble fertilizers hence nutrients are released slowly throughout the plant development. Zeolites are excellent source for this purpose. These are natural group of minerals which contains honeycomb like structure. The interconnected network contains calcium, potassium, nitrogen, phosphorous and many other trace elements. Nano membranes can be coated on to fertilizers which would ease the release and utilization of stable nutrients i.e. micronutrients, amino-acids, nitrogen, potassium and phosphorous in grain crops (Tarafdar et al. 2012a, c). Carbon nanotubes (CNT) can act as carrier which transports desired nutrients or biocides into the seeds during germination. In the same way, triazophos in a nano-emulsion could be preserved efficiently from hydrolysis in the acidic and neutral medium (Tarafdar et al. 2012b). Nanoparticles of titanium dioxide reported to enhance the plant development, photosynthetic rate and yield by 30%. Decrease in acute diseases in plants was also observed. For instance titanium nanoparticles were found to be exceptionally efficient for decreasing the severity of bacterial leaf blight and curvularia leaf spot diseases in the maize. TiO_2 nanoparticles also found to be prevalent for the occurrence of rice blast and tomato spray mold. They also have growth promoting effect in the plants (Zheng et al. 2005).

8.7 Advantages of Nanofertilizers Over Conventional Chemical Fertilizers

Nanofertilizers have several advantages over the traditional chemical fertilizers. Appropriate use of nanofertilizers is a significant way to achieve a high crop production and it also improvise soil properties. Some of the advantages of nanofertilizers in agriculture systems are discussed.

8.7.1 Seeds Germination and Growth Parameters of the Plant

Nanofertilizers are of significant importance as they positively affect the plant at earlier growth stages. It has been reported previously that by applying nanofertilizers, seed germination seedling growth and seed vigor improves well. The reason behind this phenomenon is that small size of nanofertilizers enables the nutrients to absorb into the seed and as a result shoot length and root length is enhanced. Nanoparticles have both stimulating and inhibitory effects on plant growth. Up to a certain level nanoparticles enhance plant growth but higher doses of nanoparticles can also negatively influence the plant (Pijls et al. 2009). Nano ZnO recorded higher peanut seeds germination percent and root growth compared to bulk zinc sulphate

(Prasad et al. 2012). Similarly nano-scale SiO_2 and TiO_2 have positive effects on plant germination like higher seed germination, shoot length, root length under nanofertilizers treatment over control or without nanofertilizer treated seeds (Mahmoodzadeh et al. 2013). Nanofertilizers increase availability of nutrient to the growing plant which increases chlorophyll formation, photosynthesis rate, dry matter production and result improve overall growth of the plant (Singh 2017). Similar results were reported that nano-TiO_2 treated seeds produced more dry weight, higher photosynthetic rate, chlorophyll-a formation compared to the control plants (Dos et al. 2011), which indicated that nanofertilizers significantly improved seed germination and overall growth of the plant.

8.7.2 *Crop Yield and Yield Parameters*

8.7.2.1 Sustainable Water Usage

Nano-hydrogels should be used to optimize water utilization and for sustainable agricultural production. These are capable of cyclic releasing of water and nutrients leading to effective water usage. Silver coated hydrogel added in soils can hold 7.5% more water than soils without hydrogel. Also, these can store rain water or irrigation water more than its own weight (Vundavalli et al. 2015). The volume of pollutants is also reduced by Bio-degradable hydrogels (Montesano et al. 2015; Magalhães et al. 2013). Hence, Nanofertilizers can be beneficial in arid environment. This is highly needed as drought is considered to be the important environmental risk for crop production (Jaleel et al. 2009).

8.7.2.2 Seeds Treatment with Nanofertilizers

By treating with nanofertilizers seed germination is enhanced and they can cope up with the environmental stresses easily (Adhikari et al. 2016). Seed endurance, seedling development and strength are also increased by the application of Nano fertilizers (Adak et al. 2016; Dehkourdi and Mosavi 2013; Khodakovskaya et al. 2009). It was reported that seeds coated with silver nano materials enhanced water absorption (Adhikari et al. 2016). Besides, seed treatment with nanoparticles also induced 90% enhancement in drought resistance (Rahimi et al. 2016). In addition, seed longevity during storage is also improved (Adak et al. 2016). These characteristics help to improve the yield parameters and environmental resilience (Dehkourdi and Mosavi 2013). Nanoparticles in the form of foliar spray enhance crop production significantly.

8.7.2.3 Pest and Disease Detection

It is known that plants are badly affected by the diseases and pests and almost 25% loss in rice and 50% loss in cotton is caused by the pests (Rai and Ingle 2012; Dhaliwal et al. 2010). For this purpose, organic biosensors like enzymes can be used for detecting these particular threats (Otles and Yalcın 2015; Perumal and Hashim 2014). Due to their size-related properties, nano-biosensors show an increase in accuracy, detection limits, sensitivity, selectivity, temporal response and reproducibility, compared to conventional biosensors (Huang et al. 2011).

They are able to detect single viruses and contaminants at the molecular level (Viswanathan et al. 2009). Therefore, nano-biosensors are very precise tools that can be used to prevent pest outbreaks and monitor soil quality, which enhances quality and quantity of yields (Prasad et al. 2014).

Nanofertilizers improve germination parameters of the seed as well as morphological characters such as no of leaves, leaf area index and plant height. Nanofertilizers also increase dry biomass, chlorophyll content, photosynthetic rate and translocation of photosynthats to different parts which causes more yield. According to Aschberger et al. (2015) Titanium oxide nanoparticles improved photosynthetic rate, chlorophyll content and dry weight of the plant as compared to the control. This improve translocation of photosynthetic products from source (leaves) to sink (grain, tuber, bulb, stem, fiber and leaves.) which results in more yield and quality parameters in nano-fetilizers treated plants as compared to traditional fertilizers treated plants. Also, higher value of yield parameters was observed under nanofertilizers treated plants compare to bulk nutrient sources. Iron content was enhanced in plant by treating with nano iron than control plant (Dehkourdi and Mosavi 2013).

8.7.3 Nutrient Usage Efficiency

Most of the plants do not easily take up the traditionally applied fertilizers and other products due to their instability and insoluble form (Arias-Estévez et al. 2008). In this case, nanofertilizers have advantage over traditional fertilizers because they are delivered on target site and have smart and efficient delivery system of nutrients to specific plant cells due to their small size. Also, they show enhanced stability in the environment, which improves the availability of nutrients to crops (Liu and Lal 2015). Smart delivery systems further enhance the delivery of nutrients and plant protection products through their ability of slow or controlled release (Kah 2015). This is shown to extend the electiveness of plant protection products from 3 to over 30 days (Adak et al. 2012). In addition, the effect of pesticides was found to be twice as strong with half the dose applied (Xiang et al. 2013).

Nano-biosensors can enhance this process even further by enabling smart delivery systems to precisely release nutrients in response to environmental triggers and biological demands (Rameshaiah et al. 2015).

Because of small size of nanoparticles and large surface area, nanofertilizers have the ability to provide more space which smoothen the different biochemical processes in the plant as a result photosynthesis improves. Certain properties like high reactivity with other compounds, high solubility in water and lesser size (<100) nm of nanofertilizers enables them to absorb by the plant from soil and leaves more easily. Penetration capacity and nutrient uptake by the plant are significantly increased due to higher surface area and small pore size of the leaves and roots of the plants. Reduction of particle size results in increased specific surface area and number of particles per unit area of a fertilizer that provide more opportunity to contact of nano-fertilizers which leads to more penetration and uptake of the nutrient (Liscano et al. 2000).

8.7.3.1 Reduced Runoff

The application of nanofertilizers in agricultural production has a strong potential to reduce runoff resulting from soil and water pollution caused by conventional fertilizers and remediate soils polluted with heavy metals (Karn et al. 2009). Up to 90% of agrochemicals are directly lost in the environment due to their uncontrolled application. Through increased efficiency and small size, nanofertilizers decrease leaching and reduce environmental nutrient loss (Kah 2015).

8.7.3.2 Remediation

Nanotechnology is also a solution for remediation of soils polluted with heavy metals to make them fertile again. It is most promising for countries like China and Africa which are suffering from severe soil pollution with heavy metals (Jackson et al. 2013; Qu et al. 2013). These nano techniques for soil remediation are said to be effective, economical and ecofriendly (Wei et al. 2013). A Previous study reported that iron nanoparticles reduced 99% of Trichloroethane which is a pesticide solvent, within a few days (Zhang 2003). Metal nanoparticles produced at large scale are being utilized as smart delivery systems as a result of technology expansion.

8.7.4 Nutritional Value and Health

Nanofertilizers provide greater role in crop production and several research studies revealed that these can be used to enhance growth, yield and quality parameters of the crop which results in better yield and quality food product for human and animal consumption. This translates into an improvement to three major areas of production. And if these nutrients are accumulated in reasonable amounts, then treated crops can be used as functional food as important source of nutrients like zinc,

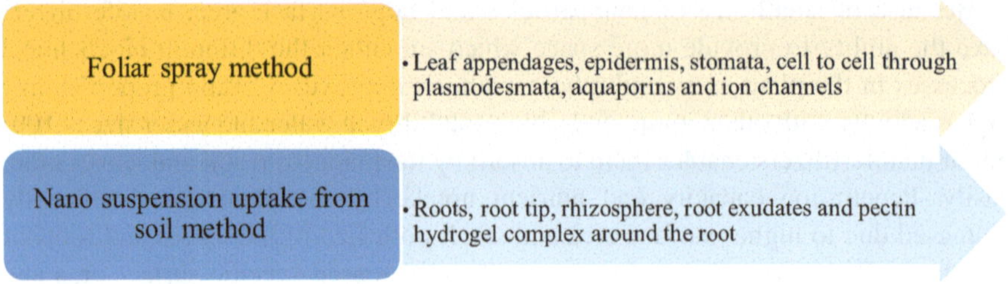

Fig. 8.3 Nanoparticles entry to plant cells

copper, iron etc. These ions are also essential for human beings and as well as animal consumption.

Researchers are taking much interest in uptake of Nanofertilizers by the plants nowadays. Nanoparticles uptake, their transfer and accumulation by plants is dependent upon the plant species, its age and the developing environment. More importantly, these procedures are related to the certain properties of the nanoparticles such as physiochemical property, functionality, and stability and delivery process. Plants take nanoparticles by the root system and transfer it to the leaves of the plant irrespective of the nutrient species (ZnO, iron, copper, aluminum and silver nanoparticles). Mostly, translocation of copper, zinc, aluminum and silver nanoparticles is observed in leaves while nickel and iron nanoparticles can be speculated in stem and cerium nanoparticles in both leaves and stem. Entrance of nanoparticles in plant cell depends upon cell wall pore size which is usually 5–20 nm. Small size of nanoparticles makes them to easily enter in the cell wall and reach up to the cell membrane. It is noticeable that pore size can be enlarged or new pores can be formed by the nanoparticles to increase their uptake by the plant. It is supposed that nanoparticles might enter into the plant system by binding with certain carrier proteins, ion channels or by endocytosis. They can also enter to the plants by root exudates and by forming membrane proteins. Nanoparticles might enter through the leaf appendages known as trichome or by stomata. A study on titanium oxide alizarin complex uptake and translocation in *Arabidopsis thaliana* seedling reported that root mucilage forms pectin hydrogel complex which is helpful for the entrance of nanoparticles complex in the plants (Fig. 8.3).

8.7.5 Reduced Nutrient Loss

To reduce the uncontrolled release of nanofertilizers in the environment, they are associated with materials such as hydrogels, films or other biopolymers such as chitosan (Kashyap et al. 2015; Hasaneen et al. 2014), which aggregate the fertilizers in complexes with mineral NPs obtained from the clay in soil or other types of ceramic materials (Choy et al. 2007) that are used for manufacturing controlled-release blocks, pots, or film for padding. These respond to environmental stimuli (such as temperature or irradiance) by modifying the release of the NFs according to the plants' need, such that more nutrients are available during times of optimal plant growth. This may reduce, for example, nitrogen losses by nitrification associated with low temperature (Derosa et al. 2010).

Another alternative for regulating NF contribution to the environment is by applying a foliar spray, especially for elements with limited bioavailability in the soil such as Fe, Cu and Ni (El-Kereti et al. 2013). Emulsions or encapsulated organic NPs can be useful for this purpose. Finally, another way to reduce the release of NPs and NMs in the environment is to match their quantities with the stage of crop growth with the greatest response.

8.8 Nanofertilizers and Nanotoxicity

With the advancement in nanotechnology various aspects of human lives have been revolutionized but there are also uncertainties associated with this technology. One of the main concerns regarding this technology is long term and unknown effects of nanoparticles on human health and environment which masks their potential benefits. To address these issues, usage of nanoparticles needs to be evaluated before complete implementation of this technology. This consideration has developed "nanotoxicology," which is responsible for assessing toxicological potential as well as promoting safe design and use of nanoparticles (Oberdörster et al. 2005). A systematic and thorough quantitative analysis regarding the potential health impacts, environmental clearance, and safe disposal of nanoparticles can lead to improvements in designing further applications of nanotechnology (Meng et al. 2009).

Even though nanoparticles are not directly related to any human diseases till now but some studies have shown their adverse effects on biological responses which may lead to toxicological consequences (Nel et al. 2006). Due to size of nanoparticles, they can pass through oral, respiratory and dermal route to animals and human beings. Now it is also assumed that due to minute size of nanoparticles, they can easily enter to cells, organelles and tissues and connect with biological structures like DNA and ribosomes. Also, size of engineered nanostructures is same to many other biological molecules such as proteins and antibodies etc. As a result, they can enter to the living systems and cause damage and serious problems to the human beings (Xia et al. 2009).

Nonetheless, it is quite disappointing that there is a lot of non-technical data also available on toxicity of nanoparticles which is giving chance to the opponents of nanotechnology to form controversial, unscientific and far reaching results about the safety of nanoparticles. This uncertainty of nanotechnology is the greatest concern of the cynics (Colvin 2003). Therefore, it is the need of the hour to form specific protocols which would assess nanoparticles outcomes in the environment and their effect in internal systems of organisms. When these problems are solved, such experimental conditions can be formed which would identify the particular nanoparticle threat to human being. For this purpose, research between multidiscipline (environmental science and material science) should be developed to overcome restrictions in identifying the threats posed by nanotechnology (Thomas and Sayre 2005).

As nanotechnology is a heterogeneous and developing subject so no single rule could be applied universally to describe its risks and control to conclude the outcome. Similarly, it is difficult to assess and compare different research groups due to lack of standard guidelines and methodologies (Dhawan et al. 2009).

Agricultural nanotechnology is also quite significant because human beings are directly affected from it (Bouwmeester et al. 2009). Nanoparticles from nanofertilizers enter to the food chain and as a result distributed to every organism in the food chain. In most of the studies, it was proposed that lower concentrations of nanoparticles have promoting effects on plant growth. Though question can be raised on their insignificant damage to health and the environment (Colvin 2003). The potential of nanotechnology in agriculture and food sectors has been identified by many countries and they are investing in its applications to food production.

It is known that various nanoparticles differ in the toxicity level. Also very limited information exists on the oral consumption of nanoparticles because most of the studies focus on toxicity of nanoparticles *in vivo* and *in vitro* system through other routes. Evaluation of nanoparticles risk should be done on individual basis. For this purpose, different factors like characterization, biotransformation, distribution, stability and elimination of nanoparticles can be involved to decide their risk assessment in the food and ultimately human beings. FDA has issued outline and implementing plan for nanotechnology and nanoparticles related products to check the scientific problems in the tools and methods for assessing these products (Chaudhry and Castle 2011).

8.9 Conclusion

Nano materials as nutrient carrier which are being used for the fabrication of nanofertilizers can greatly improvise the nutrient use efficiency in a balanced way by eliminating the environmental risk. These nanofertilizers release the required nutrients in a controlled and précised manner as per need of the plants. Nanofertilizers supply all the required nutrient elements to plants and improve their growth and development; on the other hand these also act as soil conditioning agents, making soil more fertile and productive for the cultivation. These are delivered to plants via *in vitro* or *in vivo* methods, though their translocation and mode of action in plants is still unclear. Therefore, there is a great need to study the potential mechanism of nanofertilizers in plant cells which is in fact responsible for the enhanced plant growth. On the other hand several safety concerns must be addressed prior the release of these nanofertilizers in the agricultural systems. Nano toxicological evaluation of nanofertilizers regarding the harmful impacts on human beings and environment will provide new basis for the synthesis of highly active nanofertilizers.

References

Adak T, Kumar J, Dey D, Shakil N, Walia S (2012) Residue and bio-efficacy evaluation of controlled release formulations of imidacloprid against pests in soybean (glycine max). J Environ Sci Health B 47(3):226–231

Adak T, Kumar J, Shakil NA, Pandey S (2016) Role of nano-range amphiphilic polymers in seed quality enhancement of soybean and imidacloprid retention capacity on seed coatings. J Sci Food Agric 96(13):4351–4357

Adhikari T, Kundu S, Rao AS (2016) Zinc delivery to plants through seed coating with nano-zinc oxide particles. J Plant Nutr 39(1):136–146

Ahmed O, Sumalatha G, Muhamad AN (2010) Use of zeolite in maize (*Zea mays*) cultivation on nitrogen, potassium and phosphorus uptake and use efficiency. Int J Phys Sci 5(15):2393–2401

Allen ER (1991) Supplying nitrogen, phosphorus, and potassium to plants through dissolution and ion exchange using a zeolite-based substrate. Texas A & M University, Texas, AZ

Amon M, Dobeic M, Sneath RW, Phillips VR, Misselbrook TH, Pain BF (1997) A farm-scale study on the use of clinoptilolite zeolite and De-Odorase® for reducing odour and ammonia emissions from broiler houses. Bioresour Technol 61(3):229–237

Arias-Estévez M, López-Periago E, Martínez-Carballo E, Simal-Gándara J, Mejuto JC, García-Río L (2008) The mobility and degradation of pesticides in soils and the pollution of groundwater resources. Agric Ecosyst Environ 123(4):247–260

Arthanareeswaran G, Devi TS, Raajenthiren M (2008) Effect of silica particles on cellulose acetate blend ultrafiltration membranes: Part I. Sep Purif Technol 64(1):38–47

Aschberger K, Gottardo S, Amenta V, Arena M, Moniz FB, Bouwmeester H, Brandhoff P, Mech A, Pesudo LQ, Rauscher H (2015) Nanomaterials in food-current and future applications and regulatory aspects. Int J Phys A Conf Ser 617:012032

Ayan S, Yahyaoglu Z, Gercek V, Şahin A (2005) Utilization of zeolite as a substrate for containerized oriental spruce (*Picea orientalis* L.(Link.)) seedlings propagation. Paper presented at the International Symposium on Growing Media 779

Bagdasarov V, Kazachenko A, Rustambekov M, Uspenskij B, Kuznetsova V, Efremov E (2004) Prolonged-activity nitrogen-zeolite fertilizer. Russ: RU 2222514, C2

Barbarick K, Lai T, Eberl D (1990) Exchange fertilizer (phosphate rock plus ammonium-zeolite) effects on sorghum-sudangrass. Soil Sci Soc Am J 54(3):911–916

Bouwmeester H, Dekkers S, Noordam MY, Hagens WI, Bulder AS, De Heer C, Ten Voorde SE, Wijnhoven SW, Marvin HJ, Sips AJ (2009) Review of health safety aspects of nanotechnologies in food production. Regul Toxicol Pharmacol 53(1):52–62

Brady N, Well R (1999) The nature and properties of soils. Prentice Hill, Upper Saddle River, NJ

Brunner TJ, Wick P, Manser P, Spohn P, Grass RN, Limbach LK, Stark WJ (2006) *In vitro* cytotoxicity of oxide nanoparticles: comparison to asbestos, silica, and the effect of particle solubility. Environ Sci Tech 40(14):4374–4381

Burriesci N, Valente S, Ottana R, Cimino G, Zipelli C (1984) Utilization of zeolites in spinach growing. Zeolites 4(1):5–8

Butorac A, Filipan T, Basic F, Butorac J, Mesic M, Kisic I (2002) Crop response to the application of special natural amendments based on zeolite tuff. Rostlinna Vyroba 48(3):118–124

Cengeloglu Y, Tor A, Arslan G, Ersoz M, Gezgin S (2007) Removal of boron from aqueous solution by using neutralized red mud. J Hazard Mater 142(1–2):412–417

Chaudhry Q, Castle L (2011) Food applications of nanotechnologies: an overview of opportunities and challenges for developing countries. Trends Food Sci Technol 22(11):595–603

Chinnamuthu C, Boopathi PM (2009) Nanotechnology and agroecosystem. Madras Agric J 96(1/6):17–31

Choy JH, Choi SJ, Oh JM, Park T (2007) Clay minerals and layered double hydroxides for novel biological applications. Appl Clay Sci 36(1–3):122–132

Colvin VL (2003) The potential environmental impact of engineered nanomaterials. Nat Biotechnol 21(10):1166

Dehkourdi EH, Mosavi M (2013) Effect of anatase nanoparticles (TiO_2) on parsley seed germination (*Petroselinum crispum*) *in vitro*. Biol Trace Elem Res 155(2):283–286

DeRosa MC, Monreal C, Schnitzer M, Walsh R, Sultan Y (2010) Nanotechnology in fertilizers. Nat Nanotechnol 5(2):91

Dhaliwal G, Vikas J, Dhawan A (2010) Insect pest problems and crop losses: changing trends. Ind J Ecol 37(1):1–7

Dhawan A, Sharma V, Parmar D (2009) Nanomaterials: a challenge for toxicologists. Nanotoxicology 3(1):1–9

Dikshit A, Shukla SK, Mishra RK (2013) Exploring nanomaterials with PGPR in current agricultural scenario: PGPR with special reference to nanomaterials. Lap Lambert Academic Publishing, Saarbrücken

Dos SSM, Cocenza DS, Grillo R, De Melo NFS, Tonello PS, De Oliveira LC, Cassimiro DL, Rosa AH, Fraceto LF (2011) Paraquat-loaded alginate/chitosan nanoparticles: preparation, characterization and soil sorption studies. J Hazard Mater 190(1–3):366–374

Dutta J, Sugunan A (2004) Colloidal self-organization for nanoelectronics. Paper presented at the Semiconductor Electronics, 2004. ICSE 2004. IEEE International Conference

Dwairi I (1998) Renewable, controlled and environmentally safe phosphorus release in soils from mixtures of NH 4+-phillipsite tuff and phosphate rocks. Environ Geol 34(4):293–296

El-Kereti M, El-Feky SA, Khater SM, Osman AY, El-Sherbini AES (2013) Zno nanofertilizer and he ne laser irradiation for promoting growth and yield of sweet basil plant. Recent Pat Food Nutr Agric 5(3):169–181

Flanigen E, Mumpton F (1981) Commercial properties of natural zeolites. Mineralogy and geology of natural zeolites. In: Reviews in mineralogy, vol 4. Mineralogical Society of America, Washington, DC, pp 165–175

Ghaly AE (2009) The black cutworm as a potential human food. Am J Biochem Biotech 5(4):210–220

González-Melendi P, Fernández-Pacheco R, Coronado MJ, Corredor E, Testillano P, Risueño MC, Pérez-de-Luque A (2007) Nanoparticles as smart treatment-delivery systems in plants: assessment of different techniques of microscopy for their visualization in plant tissues. Ann Bot 101(1):187–195

Gupta G, Borowiec J, Okoh J (1997) Toxicity identification of poultry litter aqueous leachate. Poult Sci 76(10):1364–1367

Hasaneen M, Abdel-Aziz H, El-Bialy D, Omer AM (2014) Preparation of chitosan nanoparticles for loading with NPK fertilizer. Afr J Biotechnol 13(31):3158–3164

He Z, Pinnau I, Morisato A (2002) Nanostructured poly (4-methyl-2-pentyne)/silica hybrid membranes for gas separation. Desalination 146(1–3):11–15

Huang ZT, Petrovic AM (1994) Clinoptilolite zeolite influence on nitrate leaching and nitrogen use efficiency in simulated sand based golf greens. AGRIS 23(6):1190–1194

Huang L, Guo Y, Peng Z, Porter AL (2011) Characterising a technology development at the stage of early emerging applications: nanomaterial-enhanced biosensors. Technol Anal Strat 23(5):527–544

Jackson P, Jacobsen NR, Baun A, Birkedal R, Kühnel D, Jensen KA, Vogel U, Wallin H (2013) Bioaccumulation and ecotoxicity of carbon nanotubes. Chem Cent J 7(1):154

Jaleel CA, Manivannan P, Wahid A, Farooq M, Al-Juburi HJ, Somasundaram R, Panneerselvam R (2009) Drought stress in plants: a review on morphological characteristics and pigments composition. Int J Agric Biol 11(1):100–105

Jampílek J, Kralova K (2015) Application of nanotechnology in agriculture and food industry, its prospects and risks. Ecol Chem Eng S22(3):321–361

Jha M, Prasad A (2006) Efficacy of new inexpensive cyanobacterial biofertilizer including its shelf-life. World J Microbiol Biotechnol 22(1):73–79

Joseph T, Morrison M (2006) Nanotechnology in agriculture and food. Eur Nanotechnol Gateway. www.nanoforum.org

Kah M (2015) Nanopesticides and nanofertilizers: emerging contaminants or opportunities for risk mitigation? Front Chem 3:64

Karn B, Kuiken T, Otto M (2009) Nanotechnology and in situ remediation: a review of the benefits and potential risks. Environ Health Perspect 117(12):1813–1831

Kashyap PL, Xiang X, Heiden P (2015) Chitosan nanoparticle based delivery systems for sustainable agriculture. Int J Biol Macromol 77:36–51

Kavak D (2009) Removal of boron from aqueous solutions by batch adsorption on calcined alunite using experimental design. J Hazard Mater 163(1):308–314

Khodakovskaya M, Dervishi E, Mahmood M, Xu Y, Li Z, Watanabe F, Biris AS (2009) Carbon nanotubes are able to penetrate plant seed coat and dramatically affect seed germination and plant growth. ACS Nano 3(10):3221–3227

Kithome M, Paul J, Lavkulich L, Bomke A (1998) Kinetics of ammonium adsorption and desorption by the natural zeolite clinoptilolite. Soil Sci Soc Am J 62(3):622–629

Köse TE, Öztürk N (2008) Boron removal from aqueous solutions by ion-exchange resin: column sorption–elution studies. J Hazard Mater 152(2):744–749

Krutilina VS, Polyanskaya SM, Goncharova NA, Letchamo W (2000) Effects of zeolite and phosphogypsum on growth, photosynthesis and uptake of Sr, Ca and Cd by barley and corn seedlings. J Environ Sci Health Part A 35(1):15–29

Kumar S, Bhanjana G, Sharma A, Sarita SM, Dilbaghi N (2015) Herbicide loaded carboxymethyl cellulose nanocapsules as potential carrier in agrinanotechnology. Sci Adv Mater 7:1143–1148

Latifah O, Ahmed O, Muhamad AN (2011) Reducing ammonia loss from urea and improving soil exchangeable ammonium and available nitrate in non waterlogged soils through mixing zeolite and sago (*Metroxylon sagu*) waste water. Int J Phys Sci 6(4):866–870

Li Z, Zhang Y (2010) Use of surfactant-modified zeolite to carry and slowly release sulfate. Desalin Water Treat 21(1–3):73–78

Li JX, Wee CD, Sohn BK (2010) Growth response of hot pepper applicated with ammonium and potassium loaded zeolite. Kor J Soil Sci Fertil 43(5):741–747

Liscano JF, Wilson CE, Norman JRJ, Slaton NA (2000) Zinc availability to rice from seven granular fertilizers. In: Arkansas Agricultural Experiment Station research bulletin, vol 963. University of Arkansas, Fayetteville, AR

Liu R, Lal R (2015) Potentials of engineered nanoparticles as fertilizers for increasing agronomic productions. Sci Total Environ 514:131–139

Liu M, Liang R, Zhan F, Liu Z, Niu A (2006) Synthesis of a slow release and superabsorbent nitrogen fertilizer and its properties. Polym Adv Tech 17(6):430–438

Magalhães ASG, Almeida Neto MP, Bezerra MN, Feitosa J (2013) Superabsorbent hydrogel composite with minerals aimed at water sustainability. J Braz Chem Soc 24(2):304–313

Mahmoodzadeh H, Nabavi M, Kashefi H (2013) Effect of nanoscale titanium dioxide particles on the germination and growth of canola (*Brassica napus*). J Ornam Hortic Plants 3(1):25–32

Malakouti MJ (2008) The effect of micronutrients in ensuring efficient use of macronutrients. Turk J Agric For 32(3):15–220

Malhi S, Haderlin L, Pauly D, Johnson A (2002) Improving fertiliser use efficiency. Better Crops 86:22–25

Manik A, Subramanian K (2014) Fabrication and characterisation of nanoporous zeolite based N fertilizer. Afr J Agric Res 9(2):276–284

Manjunatha SB, Biradar DP, Aladakatti YR (2016) Nanotechnology and its applications in agriculture: a review. J Farm Sci 29(1):1–3

Mazur G, Medvid G, Gvigora I (1986) Use of natural zeolite to increase the fertilizer of coarse soils. Soviet Soil Sci 16(4):105–111

Meng H, Xia T, George S, Nel AE (2009) A predictive toxicological paradigm for the safety assessment of nanomaterials. ACS Nano 3(7):1620–1627

Ming DW, Allen ER (2001) Use of natural zeolites in agronomy, horticulture and environmental soil remediation. Rev Mineral Geochem 45(1):619–654

Miransari M (2011) Soil microbes and plant fertilization. Appl Microbiol Biotechnol 92:875–885

Mohammadi A, Hashemi M, Hosseini SM (2015) Nanoencapsulation of *Zataria multiflora* essential oil preparation and characterization with enhanced antifungal activity for controlling *Botrytis cinerea*, the causal agent of gray mould disease. Innov Food Sci Emerg Technol 28:73–80
Montesano FF, Parente A, Santamaria P, Sannino A, Serio F (2015) Biodegradable superabsorbent hydrogel increaseswater retention properties of growing media and plant growth. Agric Agric Sci Proc 4:451–458
Naderi M, Danesh-Shahraki A (2013) Nanofertilizers and their roles in sustainable agriculture. Int J Agric Crop Sci 5(19):2229
Nel A, Xia T, Mädler L, Li N (2006) Toxic potential of materials at the nanolevel. Science 311(5761):622–627
Oberdörster G, Oberdörster E, Oberdörster J (2005) Nanotoxicology: an emerging discipline evolving from studies of ultrafine particles. Environ Health Perspect 113:823–839
Ombódi A, Saigusa M (2000) Broadcast application versus band application of polyolefin-coated fertilizer on green peppers grown on andisol. J Plant Nutr 23(10):1485–1493
Otles S, Yalcın B (2015) Chapter 8. Strategic role of nanobiosensor in food: benefits and bottlenecks. In: Rai M, Ribeiro C, Mattoso L, Duran N (eds) Nanotechnologies in food and agriculture. Springer Inter Publishers, Cham
Pandey ACS, Sanjay S, Yadav SR (2010) Application of ZnO nanoparticles in influencing the growth rate of *Cicer arietinum*. J Exp Nanosci 5(6):488–497
Perrin TS, Drost DT, Boettinger JL, Norton JM (1998) Ammonium loaded clinoptilolite: a slow release nitrogen fertilizer for sweet corn. J Plant Nutr 21(3):515–530
Perumal V, Hashim U (2014) Advances in biosensors: principle, architecture and applications. J Appl Biomed 12(1):1–15
Pijls L, Ashwell M, Lambert J (2009) EURRECA–a network of excellence to align European micronutrient recommendations. Food Chem 113(3):748–753
Polat H, Vengosh A, Pankratov I, Polat M (2004) A new methodology for removal of boron from water by coal and fly ash. Desalination 164(2):173–188
Prasad T, Sudhakar P, Sreenivasulu Y, Latha P, Munaswamy V, Reddy KR, Sreeprasad T, Sajanlal P, Pradeep T (2012) Effect of nanoscale zinc oxide particles on the germination, growth and yield of peanut. J Plant Nutr 35(6):905–927
Prasad R, Kumar V, Prasad KS (2014) Nanotechnology in sustainable agriculture: present concerns and future aspects. Afr J Biotechnol 13(6):705–713
Preetha PS, Balakrishnan N (2017) A review of nano fertilizers and their use and functions in soil. Int J Curr Microbiol App Sci 6(12):3117–3133
Qu X, Alvarez PJ, Li Q (2013) Applications of nanotechnology in water and wastewater treatment. Water Res 47(12):3931–3946
Rahimi D, Kartoolinejad D, Nourmohammadi K, Naghdi R (2016) Increasing drought resistance of alnus subcordata ca mey. Seeds using a nano priming technique with multi-walled carbon nanotubes. J For Sci 62(6):269–278
Rai M, Ingle A (2012) Role of nanotechnology in agriculture with special reference to management of insect pests. Appl Microbiol Biotechnol 94(2):287–293
Raliya R (2012) Appliance of nanoparticles on plant system and associated rhizospheric microflora. Dissertation, Jai Narain Vyas University, Jodhpur, p 199
Rameshaiah G, Pallavi J, Shabnam S (2015) Nano fertilizers and nano sensors–an attempt for developing smart agriculture. Int J Eng Res Gen Sci 3(1):314–320
Rana RK, Viswanathan B (1998) Mo incorporation in MCM-41 type zeolite. Catal Lett 52(1–2):25–29
Sasson Y, Levy-Ruso G, Toledano O, Ishaaya I (2007) Nanosuspensions: emerging novel agrochemical formulations insecticides design using advanced technologies. Springer, New York, NY, pp 1–39
Schmidt R, Szakal P (2007) The application of copper and zinc containing ion-exchanged synthesised zeolite in agricultural plant growing. Nova Biotechnol 7(1):57–62

Sharmila R (2010) Nutrient release pattern of nano-fertilizer formulations. Tamil Nadu Agricultural University, Coimbatore

Sheta A, Falatah A, Al-Sewailem M, Khaled E, Sallam A (2003) Sorption characteristics of zinc and iron by natural zeolite and bentonite. Micropor Mesopor Mater 61(1–3):127–136

Shukla SK, Kumar R, Mishra RK, Pandey A, Pathak A, Zaidi M, Dikshit A (2015) Prediction and validation of gold nanoparticles (GNPs) on plant growth promoting rhizobacteria (PGPR): a step toward development of nano-biofertilizers. Nanotechnol Rev 4(5):439–448

Singh MD (2017) Nano-fertilizers is a new way to increase nutrients use efficiency in crop production. Int J Agric Sci 9(7):3831–3833

Song JQ, Fujiyama H (1996) Difference in-response of rice and tomato subjected to sodium salinization to the addition of calcium. Soil Sci Plant Nutr 42(3):503–510

Srinivasan C, Saraswathi R (2010) Nano-agriculture—carbon nanotubes enhance tomato seed germination and plant growth. Curr Sci 99(3):274–275

Subramanian K, Rahale CS (2009) Synthesis of nanofertiliser formulations for balanced nutrition. In: Proceedings of the Indian Society of Soil Science-Platinum Jubilee Celebration, 22–25 December

Supapron J, Pitayakon L, Kamalapa W, Touchamon P (2002) Effect of zeolite and chemical fertilizer on the change of phycical and chemical properties on Lat Ya soil series for sugar cane. Paper presented at the Proceedings of the 17th WCSS Symposium, August

Tarafdar J (2012) Perspectives of nanotechnological applications for crop production. NAAS News 12:8–11

Tarafdar J (2015) Nanoparticle production, characterization and its application to horticultural crops. In: Compendium of winter school on utilization of degraded land and soil through horticultural crops for improving agricultural productivity and environmental quality. NRCSS, Ajmer, pp 222–229

Tarafdar J, Agrawal A, Raliya R, Kumar P, Burman U, Kaul R (2012a) ZnO nanoparticles induced synthesis of polysaccharides and phosphatases by *Aspergillus* fungi. Adv Sci Eng Med 4(4):324–328

Tarafdar J, Raliya R, Rathore I (2012b) Microbial synthesis of phosphorous nanoparticle from tricalcium phosphate using *Aspergillus tubingensis* tfr-5. J Bionanosci 6(2):84–89

Tarafdar J, Xiang Y, Wang WN, Dong Q, Biswas P (2012c) Standardization of size, shape and concentration of nanoparticle for plant application. Appl Biol Res 14:138–144

Thomas K, Sayre P (2005) Research strategies for safety evaluation of nanomaterials, Part I: Evaluating the human health implications of exposure to nanoscale materials. Toxicol Sci 87(2):316–321

Tilman D, Knops J, Wedin D, Reich P (2002) Plant diversity and composition: effects on productivity and nutrient dynamics of experimental grasslands. Biodiversity and ecosystem functioning: synthesis and perspectives. Oxford University Press, Oxford, pp 21–35

Treacy MM, Higgins JB (2007) Collection of simulated XRD powder patterns for zeolites, 5th edn. Elsevier, Amsterdam. Revised edition

Trenkel ME (1997) Controlled-release and stabilized fertilizers in agriculture, vol 11. International Fertilizer Industry Association, Paris

Valente S, Burriesci N, Cavallaro S, Galvagno S, Zipelli C (1982) Utilization of zeolites as soil conditioner in tomato-growing. Zeolites 2(4):271–274

VanderGheynst JS, Scher H, Guo HY (2006) Design of formulations for improved biological control agent viability and sequestration during storage. Ind Biotech 2(3):213–219

Vandergheynst J, Scher H, Guo HY, Schultz D (2007) Water-in-oil emulsions that improve the storage and delivery of the biolarvacide *Lagenidium giganteum*. Biol Control 52(2):207–229

Viswanathan S, Radecka H, Radecki J (2009) Electrochemical biosensor for pesticides based on acetylcholinesterase immobilized on polyaniline deposited on vertically assembled carbon nanotubes wrapped with ssdna. Biosens Bioelectron 24(9):2772–2777

Vundavalli R, Vundavalli S, Nakka M, Rao DS (2015) Biodegradable nano-hydrogels in agricultural farming-alternative source for water resources. Procedia Mater Sci 10:548–554

Wei Q, Yang D, Fan M, Harris HG (2013) Applications of nanomaterial-based membranes in pollution control. Crit Rev Environ Sci Technol 43(22):2389–2438

Wu L, Liu M (2008) Preparation and properties of chitosan-coated NPK compound fertilizer with controlled-release and water-retention. Carbohydr Polym 72(2):240–247

Wu S, Cao Z, Li Z, Cheung K, Wong M (2005) Effects of biofertilizer containing N-fixer, P and K solubilizers and AM fungi on maize growth: a greenhouse trial. Geoderma 125(1–2):155–166

Xia T, Li N, Nel AE (2009) Potential health impact of nanoparticles. Annu Rev Public Health 30:137–150

Xiang C, Taylor AG, Hinestroza JP, Frey MW (2013) Controlled release of nonionic compounds from poly (lactic acid)/cellulose nanocrystal nanocomposite fibers. J Appl Polym Sci 127(1):79–86

Xiao Q, Zhang F, Wang Y, Zhang J, Zhang S (2008) Effects of slow/controlled release fertilizers felted and coated by nano-materials on crop yield and quality. Plant Nutr Fertil Sci 14(5):951–955

Zhang WX (2003) Nanoscale iron particles for environmental remediation: an overview. J Nanopart Res 5(3–4):323–332

Zheng L, Hong F, Lu S, Liu C (2005) Effect of nano-TiO2 on strength of naturally aged seeds and growth of spinach. Biol Trace Elem Res 104(1):83–91

Zhou J, Huang P (2007) Kinetics of potassium release from illite as influenced by different phosphates. Geoderma 138(3–4):221–228

Chapter 9
Nanotechnology and Waste Water Treatment

Beenish Zia Butt

9.1 Introduction

Water is one of the most important and infinite reservoir from all of the renewable resources. Being essential and natural compound, water is a vital primary source for the existence of life on the planet earth. Naturally occurring water except sea water and brackish water is called as fresh water; it includes various above and underground water resources. Fresh water contributes only 3% of the earth's water and approximately one third of this is available for human consumption, while the rest is erratically circulates and polluted with domestic and industrial wastes (Patil 2015).

Contamination of water reservoirs as a result of anthropogenic activities causes water pollution. Fresh water pollution through numerous organic and inorganic pollutants causes devastating impact on environment and human health. In order to access the clean portable water several biological, physical and chemical methods have been developed including biosorption, adsorption, evaporation, ion exchange, precipitation, electrochemical treatments, flotation, reverse osmosis, membrane filtration and oxidation which are being widely used for the treatment of waste water. Though, adsorption is amongst the most productive conventional process that is being used to eliminate the noxious microorganisms and metal contaminants from water (Samanta et al. 2016; Patil 2015).

As the worlds water resources are declining day by day, so there is an emergent demand to optimize the use of water by appropriate planning and development therefore access to pure fresh drinking water is crucial to improve the quality of life by reducing the frequent disease incidence (Patil 2015). Currently, there is a frightful situation in most of the developing countries which are facing drastic shortage of drinking water. To ensure the availability and access to fresh portable water is becoming a difficult challenge as the over exploitation of water is continuously

B. Z. Butt (✉)
Monash University, Subang Jaya, Malaysia

S. Javad (ed.), *Nanoagronomy*, https://doi.org/10.1007/978-3-030-41275-3_9

causing a gradual depletion of fresh water resources (EPA 1999; Sandia National Laboratories 2003).

In many developing countries including Pakistan water scarcity has become a major issue which is needed to be addressed seriously to avoid the crises linked with water shortage. Furthermore, earth's natural water resources are being contaminated through different industrial processes and agricultural practices; these are mainly responsible for the fresh water pollution (EPA 2009). In a recent study it has been reported in India that 80% of disease incidences are associated with contaminated water containing microbial contaminants. According to World Health Organization, there should be a zero count of fecal and coliform in drinking water (Samanta et al. 2016).

Over the years, the actual necessity of water has raised due to a great outburst and rise in population and industrialization. Considerable efforts have been made to collect and reduce the over exploitation of water by generating underground water reservoirs. There are several other alternative approaches which are found to be costly, hence there is an urgent need to develop and investigate the more appropriate and cost effective methods to recycle and treat water (Qu et al. 2013). In this regard, nanotechnology is considered as an important and much functional technique that is used to improvise the quality of water (EPA 2009).

New horizons for waste water treatment include use of nano-scaled materials like magnetic nanoparticle, nano-structured iron zeolite and nano-filtration membranes which have a significant potential to remove contaminants from water. These can be used to purify water by removing chemicals, heavy metals, sediments and microbes (Patil 2015). Recycling of waste water and its management through advanced nanotechnological approaches would be helpful in reducing the environmental pollution that is becoming a foremost challenge of the present century (Pandey and Khare 2016). Nanomaterials including bioactive nanoparticles, Nano sorbents and catalytic membranes etc. are highly reactive as they have extremely small size and low surface area to volume ratio then the bulk material. Therefore these exhibit significant reactivity towards the target contaminant thus possess several advantageous over the traditional methods (Chorawalaa and Mehta 2015; El Rahman and Gepreel 2013).

Nano scaled materials are frequently used to clean the contaminated water having a large proportion of heavy metals (like lead, zinc, lead, nickel and cadmium) which are the main pollutants of the water and have detrimental effects on living beings (Qu et al. 2012). Currently the treatment of water and waste water is being carried out through nanomaterials like nano composites, metal nanoparticles and carbon nano tubes (Rao et al. 2007). Exploitation of magnetic nanoparticles as adsorbing agents presents an efficient method that can be used to elicit and remove the pollutants by providing external magnetic fields. Therefore, the potential application and use of these nanomaterials is based on the stability of nanoparticles that can be achieved by surface modification (Samanta et al. 2016).

Recent advancements in nanotechnology present significant alternative substitutes to develop new means of water supplies that will be available in future. In most of the developing countries water recycling, its supply and discharge at consumer sites greatly depends on delivery and centralized systems. The extremely effective, multifaceted and modular techniques allowed by nanotechnology are anticipated to offer highly productive and economical methods to clean waste water without the involvement of advanced infrastructure (Qu et al. 2013).

9.2 Nanotechnological Interventions in Water Treatment

Nanotechnology has facilitated the innovative means for the waste water remediation and it also assures to overwhelm the major problems coupled with the traditional methods.

There are four main nano based materials which are used in water treatment and purification (Fig. 9.1).

- Nano-adsorbent
- Nano-metals and metal oxides
- Nano-membranes
- Nano-Photocatalysis

9.2.1 Nanoadsorbents

Adsorption is a physico-chemical process that has been used since a long time to treat the contaminated and polluted water. Treatment of waste water through adsorption is mainly based on the adhesive capacity of the adsorbent which attracts the pollutants into its surface. Mainly it is a physical process occurs due to physical attractions. But chemical forces are also considered accountable for adsorption (Faust and Aly 2018). The actual phenomenon of adsorption depends on some important factors which facilitate the process. These include temperature, chemical nature of adsorbent and adsorbate, pressure and a variety of pollutants. On the other hand experimental conditions including particle size, pH, contaminants

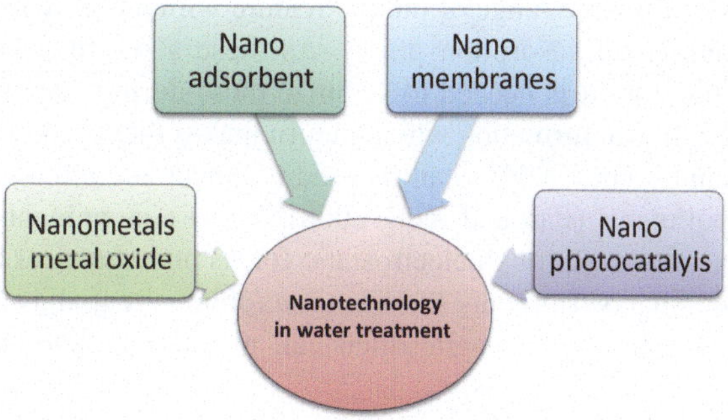

Fig. 9.1 Nanotechnological interventions in water treatment

concentration and contact time are also responsible for the adsorption of contaminants. As the adsorbent adsorbs all the maximum amount of contaminants present in the water, an equilibrium point is reached, at this point a correlation is established between the amount of contaminant adsorbed as well as in water, this is referred as adsorption isotherm. These adsorption isotherms are used to calculate the various parameters involved in adsorption process (Condon 2006; Lewinsky 2007).

Nanomaterials are found to be excellent adsorbing agents. These are considered as magical tools owing to the several unique properties like extremely small dimensions, reactivity, surface area and morphology and the presence of numerous active sites that make contact with the target contaminants specifically and help in the removal of pollutants. There are different types of metallic nanoparticles including iron and iron oxide, gold, copper oxide, nickel oxide, zinc oxide, titanium oxide and many others are being efficiently used in the recycling and purification of waste water. Adsorption technique is basically exploited for the removal and separation of inorganic pollutants mainly heavy metals. Adsorbents are the substances that adsorb the pollutants onto their surface and the pollutants that are being absorbed by adsorbents are known as adsorbate. Nano-adsorbents are novel, most potent and have a significantly high rate of adsorption for the elimination of organic and inorganic pollutants. In order to save the adsorbent materials, the higher process efficiency facilitates the application of condensed wastewater decontamination materials with extremely smaller footmarks, mainly for decentralized systems and applications.

There are mainly three types of carbon nano-adsorbents which have remarkable adsorption capacity (Gehrke et al. 2015).

1. Carbon nanotubes (CNTs)
2. Polymeric nano-adsorbents
3. Zeolites.

9.2.1.1 Carbon Nanotubes

Carbon based nano-adsorbents are amongst the extensively studied nanomaterias used for the decontamination of wastewater (Fig. 9.2). Carbon nanotubes (CNTs) are cylindrical modified structural forms of carbon. These are of two types of CNTs, one is single walled and the other is multi-walled, based on their fabrication. CNTs show a wide range of unique feature with a highly reactive surface chemistry with numerous adsorption points. As these show hydrophobic surface characteristics therefore it is needed to stabilize these during adsorption process to minimize the aggregate formation which in turn makes the reactive sites unavailable for contaminants. These CNTs can be used to detect and eliminate several non biodegradable pollutants (Pan and Xing 2008). Metal contaminants are absorbed by the carbon nanotubes through electrostatic forces and chemical bonding. Moreover these carbon nanoadsorbents show antimicrobial potential as these cause oxidative stress in microbes by damaging their cell membranes, though in this process no noxious

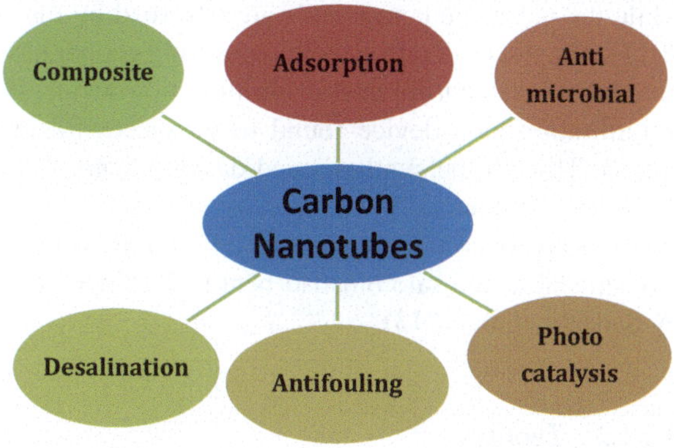

Fig. 9.2 Application of carbon nanotubes in water treatment

Table 9.1 Use of CNTs in the removal of water pollutants by mechanism (Usmania et al. 2017)

Sr. No	Water pollutants	Carbon nanotubes as adsorbents	Metal removing capacity
1	Co(II)	Multi walled carbon nanotubes (MWCNTs)	69.63 mg/g
2	Cd(II)	MWCNTs	75.84 mg/g
3	Pb(II)	MWCNTs	101.05 mg/g
4	Zn(II)	MWCNTs	58 mg/g
5	Cu(II)	MWCNTs	50.37 mg/g
6	Cr(VI)	MWCNTs–AC	9.0 mg/g
7	Ni(II)	MWCNTs/iron oxides	91.8 mg/g
8	As(III)	MWCNTs	0.08 mL^{-1}

chemical compounds are produced. Therefore it is considered as an edge over the traditional methods (Liu et al. 2013).

Through apposite alteration during operating process can be used to re-synthesize these nanomaterials. In a recent study Yang et al. reported the synthesis of plasma-modified ultra-long CNTs which exhibit higher adsorption properties for salt removal then the usual decontamination methods (Yang et al. 2013).

These are the most promising nanomaterials which have great implementations in water purification for future generations with significant filtration, desalination and disinfection features (Table 9.1). According to a recent report, a sponge equipped with CNTs and boron has fabricated by a group of US researchers which exhibits an excellent potential to remove oil from water (Hashim et al. 2012). Despite of several advantages their exploitation on large scale to treat municipal wastewater is intractable due to hefty development costs. Though, these are much more effective in point-of-use applications (De Volder et al. 2013).

9.2.1.2 Polymeric Nano-adsorbents

Polymeric nano-adsorbents are another class of adsorbents, these include dendrimers which are highly branched molecules used to remove organic and inorganic substances. They possess inner hydrophobic core that adsorbs organic pollutants while the inorganic compounds are adsorbed by the outer branches (Khajeh et al. 2013). The use of dendrimers in recovering copper has been demonstrated by incorporating these dendrimers in an ultrafiltration device. The combination of dendrimer-ultrafiltration as a device found to be successful in removing large amounts of copper. Though the synthesis and development of dendrimers is multifaceted and these are commercially supplied by only few companies of China (Diallo et al. 2005). To recover anionic substances from industrial effluents a remarkable potent biodegradable and safe bioadsorbent is developed by coupling chitosan dendrimer by Sadeghi et al. (2013).

9.2.1.3 Zeolites

Zeolites are aluminosilicate, microporous minerals frequently used as adsorbing agents. Silver embedded zeolites are most important nanomaterials and have been acknowledged since 1980s. As, these nanadsorbents have pores which can be incorporated by nanoparticles, where these are replaced by other ions (Nagy et al. 2011).

The Water Research Commission Report No KV 297/12 states the exploitation of zeolites in combination with silver nanoparticles to be used as adsorbing agents as a disinfectant means (Petrik et al. 2012). Tiwari et al. (2008) and Jung et al. (2004) developed and used nanozeolites in waste water treatment (Table 9.2).

9.2.2 Nano-metals and Nano-metal Oxides

Nanoscale metals and metal oxides are most potent means to recover metal contaminants and considered as substitutes to adsorbing materials. There are few nano-metal oxides which facilitate the removal of metals using a low gradient magnetic field. Nano iron hydroxide is an active adsorbent that resist abrasion, it has a high surface area which facilitates the adsorption process during arsenic removal from water (Aredes et al. 2012). Commonly, these nano-metals and their oxides are compacted bits or used in powders for industrial use (Qu et al. 2013).

9.2.2.1 Silver Nanoparticles (AgNPs)

Silver and its nanostructures are excellent biocides which are extensively used as antiseptic agents and destroy the microbial pathogens of wastewater (Jain and Pradeep 2005). These have a wide range of applications, though these are extensively used as anti-microbial, anti-fungal, anti-viral and also used as decontaminating agents in water purifying process (Masel and Masel 1996). Nanosilver is well known for its potential antimicrobial activity which greatly depends on silver ions that released from the sheets or layers of nanosilver. Therefore, these nanosilver coatings and inner materials have to be transformed after a certain period to

Table 9.2 Removal of metal pollutants by adsorption mechanism (Usmania et al. 2017)

Sr. No	Metal pollutant	Nanoadsorbent nanoparticles	Metal removing capacity
1	As(III)	ZVI	3.5 mg/g
2	As(III)	Akaganeite	–
3	As(III)	ZVI onto activated carbon	1.997 mg/g
4	As(III)	Mt-nZVI	59.9 mg/g
5	As(III)	Iron chitosan	94 mg/g
6	As(III)	CS-NZVI-CMβ-CD	18.51 mg/g
7	As(III)	a-Fe_2O_3	2.41 mg/g
8	As(III)	Mag-Fe-Mn	47.76 mg/g
9	As(III)	Fe_2O_3	20 mg/g
10	As(III)	Fe_2O_3	188.6 mg/g
11	As(V)	y-Fe_2OOH	37.3 mg/g
12	Cd(II)	Thiolactic acid modified TiO_2	–
13	Cd(II)	Cu I	136.3 mg/g
14	Cd(II)	Akaganeite (2.6 nm)	17.1 mg/g
15	Cd(II)	Alumina–silica (32 nm)	
16	Cd(II)	S-nZVI	83.19%
17	Co(II)	S-nZVI	73.82%
18	Co(II)	nZVI	59.63%
19	Cr(VI)	Akaganeite (3–6 nm)	80 mg/g
20	Cr(VI)	$MnFe_2O_4$ (10 nm)	31.5 mg/g

improvise the adsorption efficiency. But the modification cost of these nano-adsorbents is high which limits the use of nanosilver as disinfectant (Quang et al. 2013). The use of nanosilver has been reported since 1800s for the photo development process, its use in swimming pool has been certified by the Environmental Protection Agency, as algaecides and purifying agent for drinking water. Though, it has excellent and wide ranged activity against several microorganisms (Nowack et al. 2011).

9.2.2.2 Titanium Oxide Nanoparticles (TiO_2NPs)

Titanium oxide nanoparticles are chemically stable, least toxic to humans and can be used as disinfecting agents and to treat water contamination (Kim et al. 2012; Savage and Diallo 2005). These are more superior and have many advantages over the nanosilver metals particles due to their infinite life time. When used as a catalyzing agent it resists the change in its chemical composition during degradation and antimicrobial processes. Though, TiO_2 are activated through ultraviolet lights, while no energy is needed for nanosilver to exhibit its antimicrobial activity (Quang et al. 2013). Photocatalytic TiO_2 nanoparticles have a number of advantages over the other metallic nanoparticles, as these are inert, abundantly available in abundance, less expensive and have a wide range activity to degrade chemical contaminants including organic compounds and microbes. These features make it an efficient, resilient and most potent nanoadsorbent, used to treat water at small and large scales. Toxicological effects of the above described metal nanoparticles and oxides have found to be low, but the further surface modification during process may lead to toxicity so these should be re-evaluated for safety concerns after making chemical alteration (Gehrke et al. 2015).

9.2.2.3 Magnesium Oxide Nanoparticles (MgONPs)

Magnesium oxide nanoparticles and its other nanoscale structures are used to remove halogens as they have a high surface area than the bulk material. MgO-NPs generally range from 5 to 100 nm with specific surface area from 25 to 50 m^2/g. Photocatalytic nanoscale materials also offer the elimination of degradation of various types of pollutants including effluents and agrochemicals. Effectiveness of Mg- and MgO-NPs as effective bactericidal agents against different species of Gram-positive and Gram-negative bacteria (Stoimenov et al. 2002).

9.2.2.4 Gold Nanoparticles (AuNPs)

Gold nanoparticles and its other nano-structural forms including nano-powder and nanodots are spherical, brown colored metal nanoparticles that usually are 20–100 nm in diameter with a specific surface area of about 1–3 m^2/g. Palladium coated gold nanoparticles are most efficacious and active in removing organic compounds particularly tri-chloroethane (TCE) from water, palladium coated gold NPs have found to be 2200 times superior than the palladium and gold NPs alone (Brittany et al. 2006).

9.2.2.5 Zinc Oxide Nanoparticles (ZnONPs)

Zinc oxide NPs have been reported as most effective catalysts to absorb heavy metals from contaminated water specifically arsenic, although zinc oxide in bulk form is unable to remove arsenic (Brittany et al. 2006).

9.2.2.6 Aluminum Dioxide (Al_2O_3NPs)

Heavy metals elimination from water and waste water can be achieved using aluminum oxide as an adsorbent. The use of Al_2O_3NPs in comparison to Alumina as bulk material is found to be more reactive in removing metallic ions from water due to unique features of nanomaterials (Goswami et al. 2012).

9.2.2.7 Magnetic Nanoparticles

Magnetic nanoparticles have widely been studied and reported as good sorbents and have a profound application in water treatment process, to recover and remove arsenic. These can be directly introduced into the polluted ground water and later on after a period of operation can be recovered using magnetic field (Jung et al. 2004).

As magnetic nanoparticles have also been investigated and employed for various medical applications, US Food and Drug Administration have approved these as contrasting agents in magnetic resonance imaging. Though, to evaluate the toxicity of magnetic nanoparticles a substantial database has been generated that facilitate the production and use of magnetic nanoparticles with least toxic effects to remove contaminants from water.

9.2.2.8 Iron Oxide Nanoparticles (Fe_2O_3NPs)

Iron and its different mineral compounds like akaganeite, goethite, lepidocrocite, feroxyhyte, ferrihydrite, hematite, magnetite and maghemite are frequently used in decontamination processes as adsorbents. Organic substances exhibit exceptionally high rate of adsorption towards the metal nanoparticles, it has been estimated that it takes only 30 min to adsorb 90% of the organic pollutants (Brittany et al. 2006).

Ferrites are magnetic iron oxide compounds, have crystalline structure. Due to its strong magnetization property, iron atoms from iron ferrite are rapidly exchanged by other metallic ions. Therefore, different natural magnetite and ferrites are employed in wastewater treatment to recover actinides and other metals. The application of NPs of ferrite and other iron containing compounds in decontamination process have few advantages over the other methods; significantly higher rate of metal removal can be achieved as the higher surface area to volume ration of NPs can seriously improve the adsorption capacity of NPs. On the other hand, reactive sits on the surface are occupied by different functional groups which greatly facilitate the adsorption of target metal contaminant (Gehrke et al. 2015).

9.2.2.9 Nano-zero Valent Iron

It can be used as a substitute for the treatment of groundwater pollution in addition with perchlorates and chlorinated hydrocarbon. It is highly reactive than the iron particles. These zero valent iron nanoparticle have a very short life time, therefore further investigation should be carried out to make these more stable by surface alterations (Aredes et al. 2012; Matlochová et al. 2013).

9.2.3 *Nano-membranes*

Membrane technology is a highly advanced approach used for the remediation of wastewater. These membranes act as physical barrier for contaminant substances and their efficiency depends on their pore and molecule size. Recent technological developments has also increased their specificity and functionality, and made them most advanced and promising means to treat water.

There are few types of nano-membranes which are used in water treatment and purification systems.

1. Nano-filtration membranes
2. Nanocomposite membranes
3. Self-assembly membranes
4. Nanofiber membranes
5. Aquaporin-based membranes
6. Graphene membranes
7. Cellulose membranes

9.2.3.1 Nano-filtration Membranes

Membrane filtration is a pressure driven process which redistricts the entry/access of substances with a diameter of less than 0.5 nm. The charge based repulsion system is a distinctive feature of these membranes that facilitates the recovery of various metal contaminants. Nano-filtration membranes are mainly used as water softening agents, to reduce foul smell and color and to remove heavy metals from groundwater reservoirs. It is cost effective approach, can be significantly employed to desalinize water (Jagadevan et al. 2012).

9.2.3.2 Nanocomposite Membranes

Nanocomposite membranes are a novel and reliable material used in separation process. These are mainly consisting of two types of membranes including surface-functionalized membranes and mixed matrix membranes. Nano-fillers, which are mostly inorganic substances, are embedded in a matrix material of membranes. Nano-fillers have a high surface area to volume ration with specific surface area (Feng et al. 2013; Wegmann et al. 2008).

Incorporation of nano-fillers like metal oxides nanoparticles can be done to achieve thermally and mechanically stable membranes. Water permeability of these membranes can be increased by embedding zeolites which enhance membrane hydrophobicity. Silver NPs, carbon nanotubes and Titanium dioxide NPs are usually used to overcome and reduce fouling. Kim and Deng (2011) synthesized a most advanced type of nanocomposite membrane by incorporating mesoporous carbons into thin film polymeric membranes. These membranes are semipermeable having an upper selective layer that helps in reverse osmosis (Kim and Deng 2011).

9.2.3.3 Self-Assembly Membranes

These are one of the most novel and advanced techniques that facilitate the permeation of gases. Self-assembly is defined as the “autonomous organization of components into patterns or structures without human intervention” (Whitesides and Grzybowski 2002). Their structural integrity can be maintained systematically by managing different reaction parameters (Qiu et al. 2012).

These membranes have same characteristics as ultrafiltration and offer improved efficacy. Development of self-assembling membranes have been achieved on nanoscale but the synthesis of these membranes on commercial scale is not feasible due to processing complications (Pendergast et al. 2013).

9.2.3.4 Nanofiber Membranes

Electrospinning is a nano fiber production method, it employs the electric force to draw and join different charged substances and polymers. These nanofibers diameter ranges in nanometers, and are frequently used in filtration and separation processes. These can be modified by making specifications as per desired application (Wegmann et al. 2008).

Electrospinning technique is a highly established process to clean contaminated air but its exploitation in water remediation is still not clear. Therefore, there is a great need to investigate the potential aspects in electrospinning to treat wastewater. NanoCeram® is a patented small sized nanofiber with an extremely high surface area of 300–600 m^2/g, electropositive filter medium can be synthesized in large amounts. It is a white powder having fibers of 2 nm that assemble in aggregate forms. It can be used for pre-filtration processes used for ultrapure water systems, deionized water for laboratory use from industrial wastewater (Karim et al. 2009).

9.2.3.5 Aquaporin Based Membranes

Aquaporins are integral membrane proteins that form pores in biological membranes, and facilitate the passage of water molecules between and within living cells. These are also known as water channels which limits the transport of other ionic molecules. These offer high permeability for water and allow selective entry, therefore are considered as suitable nanomaterials to develop more efficient biomimetic membranes (Tang et al. 2013).

Aquaporin Inside™ (Aquaporin A/S, Copenhagen, Denmark) is the first commercial aquaporins incorporated biomimetic membrane. It can be used to treat saline water as its pressure withstanding capacity is up to 10 bar and can offer a water flux of 100 L/(h m^2), There are no such aquaporin membranes which can withstand operating pressure permanently and intense reaction conditions during the process. To commercialize aquaporin membranes for water remediation purposes, vast investigation is needed in order to improve their stability and efficiency (Xie et al. 2013).

9.2.3.6 Graphene Membrane

It is a semi metal, allotrope of carbon, the use of graphene based membranes in water remediation has not been established extensively, though it possesses many distinctive characteristics like carbon nanotubes. Therefore, the exploitation of graphene in the fabrication of nanocomposite membranes is under consideration to compete and overcome the principal limitations that are encountered during CNTs application in wastewater treatment. It has been observed that the smooth surface plane and edges are characteristic features of graphene which are readily available for chemical modification making these more reactive and increase their work efficiency (Lee et al. 2013).

Furthermore, it has been investigated that the surface oxidation of graphene and its addition in polymeric membrane make it hydrophobic and facilitate water infiltration and reduce fouling (Krishnan et al. 2008) (Fig. 9.3).

9.2.3.7 Cellulose Membrane (CM)

It is most abundant organic polymer compound on earth. Its micro and nanofibers sizes and mechanical strength can be employed in the fabrication of cellulose based membranes for water treatment processes. Some cellulose membranes have also been synthesized through immaculate cellulose nanomaterial mats and by incorporating CM in other suitable matrices like poly vinylidene fluoride (Fernandes et al. 2013; Mautner et al. 2014).

The use of polymeric cellulose membranes has also been investigated in various membrane filtration processes like microfiltration, nano-filtration and ultrafiltration etc. These seem to be more biocompatible and possess many advantages over the other methods as surface hydrophilicity alteration, improved permeability and selectivity and most active in reducing biofouling (Lalia et al. 2014).

9.2.3.8 Modification of Nano-membrane by Different Types of NPs

Alterations of nano-membranes through different types of nanoparticles can be achieved to improve their filtration efficiency.

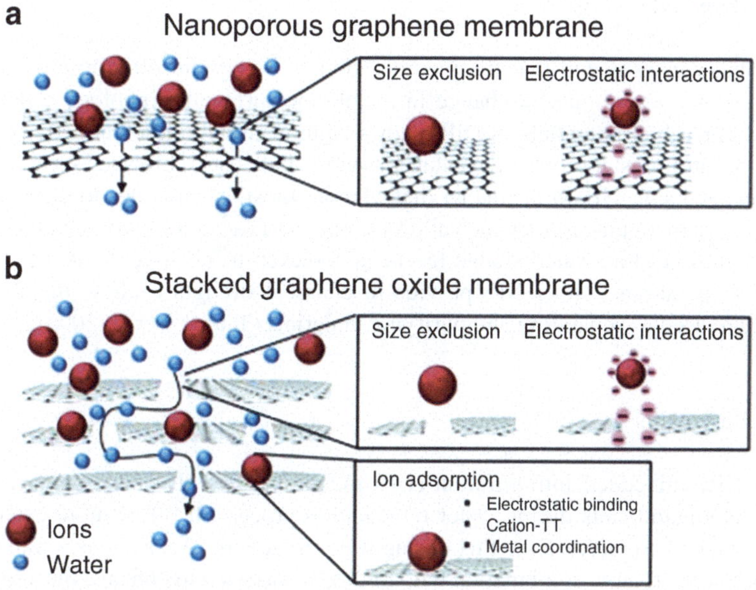

Fig. 9.3 Separation mechanism of (**a**) a monolayer graphene membrane with nanopores of controlled size (**b**) a multilayer graphene membrane composed of stacked graphene oxide sheets. (Reprinted with permission from Perreault et al. 2015, Copyright 2015 The Royal Society of Chemistry)

Titanium Based NPs

Titanium oxide (TiO_2) NPs have excellent surface characteristics thus considered appropriate nano-filler that is employed in the synthesis of Ultra Filtration (UF) polymeric membranes. The addition of NPs as nano-filler in the polymeric membranes can be used to improvise the separation efficiency (Cao et al. 2006). Moreover, the distribution and abundance of TiO_2NPs on the surface of polymeric membrane is an important factor which influences the membranes performance in general (Bae and Tak 2005; Chou et al. 2005).

NPs embedded polymeric membrane allows more surface area for contaminants interaction. Two forms of TiO_2, rutile and anatase, have structurally distinct features. Anatase (TiO_2NPs) forms pores of smaller diameter on the membrane surface and slits within membrane while rutile makes pores with larger diameter. Therefore smaller anatase can be used to enhance the antifouling and permeability characteristics of the membrane and can be effectively used in water treatment processes. TiO_2 embedded ultrafiltration membrane allows the rapid dispersal of water on the membrane surface which facilitate the photocatalytic degradation of pollutants. TiO_2 embedded ultrafiltration membrane allows the rapid dispersal of water on the membrane surface which facilitate the photocatalytic degradation of pollutants (Zhao et al. 2012).

Silver Based NPs

It has been investigated that the incorporation of AgNPs in ultrafiltration membranes brings a pronounced change in membrane surface morphology efficiency and antimicrobial potentials. AgNPs shows significant antimicrobial activity with low concentrations, it provide advantages over the bulk use of biocide as water purification agents. AgNPs with smaller diameter are most valuable due to their higher surface area to volume ratio, which allows a more surface area for microbial interaction. Filtration of activated sludge has been achieved using silver coated polysulfone UF membrane, silver NPs provide mechanical strength to these membranes which are used as separating and purifying materials (Taurozzi et al. 2008).

Silica Based NPs

Silica NPs embedded ultrafiltration (UF) membrane has also been employed to separate contaminants during water remediation process. It offers more extensive inner channels to increase the flux during water treatment (Arthanareeswaran et al. 2008). Polysulfone in combination with SiO_2NPs was used to enhance the antifouling and permeability properties of the membrane (Ahmad et al. 2011).

Zinc Oxide Based NPs

Zinc oxide NPs have well established antimicrobial potential of ultratfiltration polymeric membrane when in turn improves the antifouling resistance of these membranes. Scientists have reported the development and use of ZnONPs embedded membranes for wastewater treatment, it was observed that the incorporation of ZnO-NPs as nanofiller improves the water permeability (Langlet et al. 2006).

9.2.4 Photocatalysis

Photocatalysis is a light driven chemical oxidation reaction used in water purification process; in general it is the oxidative removal of micro contaminants and microorganisms (Friedmann et al. 2010). It has been reported that most of the organic contaminants are readily deteriorated by photocatalysis. The use of TiO_2 in this regard has been extensively studied (Table 9.3) as it is less expensive, less toxic, availability in access, and many remarkable features makes it an ideal photocatalyst (Qu et al. 2013).

Many persistent substances such as antibiotics and other mico-contaminants can be degraded photocatalytically during tertiary cleaning steps in wastewater. Ultraviolet lamps are generally used for the activation of TiO_2, though sunlight can also be used. KRONO Clean 7000 (Kronos Inc., Cranbury, NJ, USA) is a unique

Table 9.3 CNTs in the removal of water pollutants by photocatalysis mechanism (Usmania et al. 2017)

Sr. No.	Organic pollutants	Carbon nanotubes catalyst
1	Rhodamine B	CNT/TiO_2
2	Cango Red	$SWCNT/TiO_2$
3	Eosin Yellow	$MWCNT/TiO_2$
4	Acid Blue 92	$MWCNT/TiO_2$
5	Aniline, nitrobenzene, benzoic acid	CNT/TiO_2
6	Methyl Orange	CNT/TiO_2
7	4-Nitrophenol, Rhodamine B	CNT/titanium silicate
8	Acetaldehyde	Titania nanotube/CNT
9	Methyl Orange (MO)	$MWCNT/TiO_2$
10	*Bacillus cereus* (*B. cereus*)	MWCNTs–TiO_2

photocatalyst, allow the use of sunlight. In addition to TiO_2, tungsten trioxide has a photocatalytic effect when exposed to visible light (Meng et al. 2013).

In past few years several researchers have also described the separation and catalytic processes in order to clean wastewater. The use of most productive nanoparticles in filtration process like nano-filtration has to be applied to limit the entry of toxic materials. Soil particles and most microbial pathogens, viruses, spores and other pollutants are trapped and degraded by microfiltration membranes. Recently, some advanced oxidation techniques were effectively established commercially to treat wastewater. An accredit company, Purific Water (Holiday, FL, USA), has developed water purification and softening treatment by combining ceramic membrane and photocatalysis, it has a great capacity to treat water by degrading highly toxic compounds including dioxine and volatile organic substances (Azrague et al. 2007; Keuter and Gehrke 2012).

9.3 Applications of Nanotechnology in Wastewater Remediation/Filtration

There are several organic and inorganic pollutants which can be removed and degraded by the application of different nano materials.

9.3.1 Removal of Metallic Pollutants

Heavy metals are naturally occurring trace elements with toxic effects to all types of living forms. These metal pollutants enter in the human body through consumption of contaminated water and food which causes severe lethal effects on the human body. These are non biodegradable that persist in environment for a long period of time. Different chemical and physical methods are employed for the recovery and removal of heavy metals but these techniques are least productive. Thus there is a need to develop a new most efficient approach for the recovery of heavy metal pollutants. The use of nanotechnological innovation has also provided a more advanced means to treat wastewater for the removal of heavy metals contaminants. Adsorption is a conventional method that is being used for the removal of organic and inorganic contaminants, the use of nanomaterials as adsorbents seem to be an advanced method for water remediation (Mohammed et al. 2011).

9.3.1.1 Arsenic

It is one of the most concerned metal contaminant of drinking water, as the groundwater reservoirs are heavily contaminated with arsenic and it is a foremost challenge for developing countries (Brammer and Ravenscroft 2009). Different technologies have been used for arsenic elimination from water, and the efficiency of these technologies relies on the physical and chemical properties of the arsenic compounds present in water. Among all the technologies applied for arsenic removal (coagulation-filtration, membrane separation, ion exchange and adsorption), adsorptive separation provides several benefits over others that including low operation cost, simple and stable operation, compact facilities, the absence of added reagents, and easy handling of waste (Akin et al. 2012). Adsorption technique utilizing NPs provide significant improvement with their high specific surface area and sorption sites (Qu et al. 2013).

9.3.1.2 Cadmium

It is a naturally occurring soft, whitish blue colored toxic heavy metal, that is chemically similar to zinc and mercury. All compounds of cadmium are highly toxic in nature and it has been listed as a carcinogen like other heavy metals by the International Agency for Research on Cancer. Due to its potential harmful effects the use of cadmium is mainly controlled by Environmental Protection Agency (EPA) in the United States (US). The contamination of cadmium in drinking water causes severe health effects diarrhea, vomiting, liver disease, high blood pressure, lung damage, cancer and several others (Ostad and Wise 2005).

Skubal et al. (2002) reported the use of titanium dioxide NPs as nanoadsorbent to recover and separate cadmium from contaminated water. Results showed that the use of nanmaterials as adsorbents could be a promising way to remove cadmium from water. The elimination of cadmium from water has been investigated and reported using alumina silica NPs and anatase nanoparticle as water purifying agents (Gao et al. 2004; Pacheco et al. 2006).

9.3.1.3 Chromium

Chromium is a lustrous, hard and brittle metal which causes environmental pollution including soil and water pollution. It occurs in nature in different form but the trivalent and hexavalent chromium are most commonly occurring forms. In comparison to other forms, trivalent chromium is less toxic to human. Though, hexavalent chromium compounds are deadly toxic to animals and human and also a major cause of cancer. Due to rapid industrialization and anthropogenic activities the concentration of chromium in the soil has increased to a considerable extent. Cr(VI) is water soluble and leached easily from soil to groundwater where it causes water contamination (Goodman 2011).

Nanofiltration membrane has also been reported for the elimination of Cr(VI) from contaminated water. Adsorption of Cr(VI) using FeO nanoparticles, from ground water resources has also studied (Ren et al. 2010; Shao-feng et al. 2005).

9.3.1.4 Cobalt

It is a chemical element of earth crust, its consumption in higher concentration could cause potential toxic effects, and International Agency for Research on Cancer has established the cobalt as a potential carcinogen. Various industrial processes (metallurgical, electroplating and petrochemical paints etc.) are contributing the higher amounts of cobalt in soil and water.

9.3.1.5 Copper

It is widely used as agrochemicals in agriculture to control the mushroom, and as algicides in the remediation of water. Excessive intake of copper can result several health ailments. According to EPA it is not carcinogenic and permissible level of copper in drinking water is 1.3 mg/L. Different agricultural practices and industrial processes are major causes of copper contamination in surface and ground water (Deliyanni et al. 2003; Sun et al. 2006). Banerjee and Chen (2007) reported the synthesis and use of magnetic NPs as adsorbent for the removal of a copper from polluted water.

9.3.2 Removal of Dyes

These are the main pollutants produced by industries textile, leather, paints, printing, rubber and paper. Dyes are mainly separated and eliminated through the use of nano-adsorbents. TiO_2NPs has been employed for the removal of red 195 azo dye.

Similarly, the use of Fe_3O_4, silica nanosheets, chitosan NPs has also been investigated for the removal of orange G, methylene blue, azo dyes respectively from wastewater (Chang and Chen 2005; Chen et al. 2011).

9.3.3 *Removal of Pesticides*

The main sources of agrochemicals are agriculture practices including pesticides, chemical fertilizers etc. The extensive application of these chemicals is polluting the environment. Use of nano materials to eliminate and degrade chemical pesticides is still under investigation, there are only few reported studies (Ali and Jain 1998).

9.3.4 *Removal of Microorganisms*

Microbes are the other main contaminants of water which are responsible for water contamination and cause several diseases. The exploitation of nanomaterials to degrade and remove microbial contaminants from water has been extensively studied (Li et al. 2010; Shen et al. 2010).

9.4 Nanomaterials as Disinfecting Agents in Water Treatment

Antimicrobial properties of nanomaterials have also been investigated against several pathogenic microorganisms. Microbial water contamination is one of the main concerns of portable water, though nanomaterials are frequently used as antimicrobial agents in water treatment and purification, proficiently inhibit the microbial growth and act as disinfecting agent. Though, the exploitation of antimicrobial nanomaterials has also revealed the formation of some toxic by-products (Li et al. 2010) (Fig. 9.4).

Most of the reported DBPs are carcinogenic in nature and have lethal effects on human health (Krasner et al. 2006; Hossain et al. 2014). The other most commonly used method is UV-disinfection method, it generates less DBPs but it is not much efficient for viruses. All these limitations associated with the conventional methods support the development and use of other most suitable methods. Appropriate disinfecting agents should have following characteristics; it should have wide range antimicrobial potential, highly soluble, safe for humans, cheap and simple in application and preservation at ambient conditions. Nanomaterials including silver, zinc, titanium and carbon nanotubes have showed excellent properties against microorganisms (Li et al. 2008; Hossain et al. 2014).

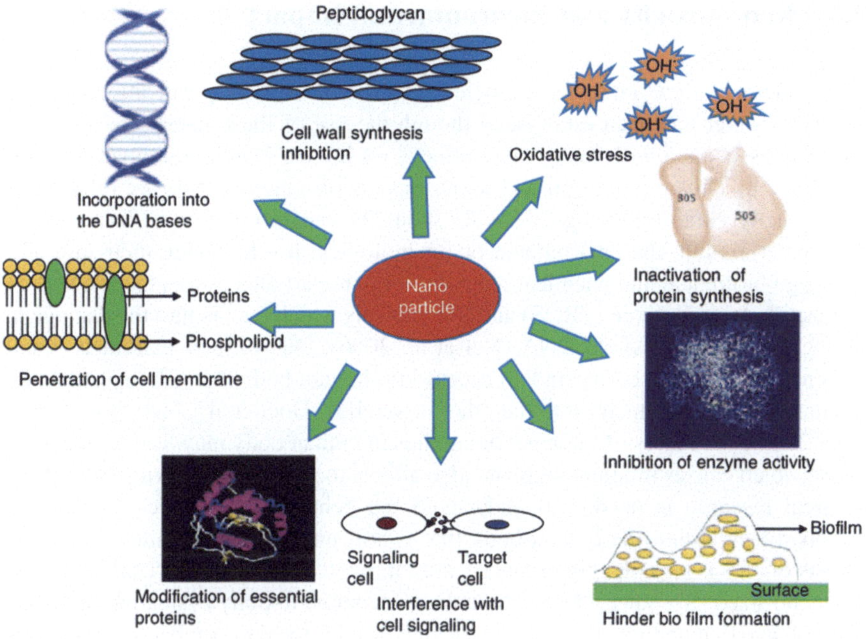

Fig. 9.4 Mechanisms of toxicity of nanoparticles (NPs) against bacteria (Singh et al. 2014)

9.4.1 Titanium Dioxide NPs as Antimicrobial Agents

Titanium dioxide nanoparticles are one of the most important nano materials used in the treatment of water (Friedmann et al. 2010). Production of reactive oxygen species results from the titanium oxide nanoparticles is responsible for their remarkable antimicrobial activity; ROS cause the destruction of microbial cells (Hossain et al. 2014).

9.4.2 Silver NPs as Antimicrobial and Disinfectant Agents

Antimicrobial action of silver nanoparticles has a well-known usage from decades ago. Currently nano scale silver has been reported for its extensive use as antimicrobial agent. These are considered as the most appropriate disinfecting agents used in water treatment to destroy the microbial contamination. Silver ions that are released from nano sized silver materials get attached and destroy the cell membrane, DNA and most important enzymes which regulate the cell metabolism and hence destroy the microorganisms (Qu et al. 2013). Carbon nanomaterials (carbon nano tubes and graphene) have also been reported for their outstanding antimicrobial activities, CNTs have found more efficient in removing microbial contaminations (Vecitis et al. 2010).

9.5 Nano-toxicity and Environmental Impact

The implementation of nanotechnological advancements in the purification of water and wastewater has been established though the use of these nanoscale materials and it is associated with many toxic effects, on human health and environmental degradation. Thus the harmful and toxic impacts of nanomaterials are referred as nanotoxicity. Nano-toxicological studies should be carried out prior to the release of nanomaterials into the environment or for biological use to ensure their safe use. The high biological and chemical reactivity of nanoparticles causes the production of reactive oxygen species (ROS) and free radicals which results into the damage to biological membranes and DNA (Nel et al. 2006). Due to their extremely small dimensions, NPs can easily find an access into human body through ingestion and inhalation and can transfer into the other organelles (Hoet et al. 2004; Oberdörster et al. 2005). An excess of these nanomaterials in human body may lead to oxidative stress which causes inflammation and also affects the immunity (Savić et al. 2003). A great research is needed to understand the behavior and mode of action of nanoparticles within the human body that is still unknown. To evaluate the toxic effects of NPs is a challenging issue because it involves the characterization of various nano sized substances. Hence to make an accurate toxicity evaluation of NPs is difficult and demanding. These issues need to be addressed to combat the hazardous effects of nanomaterials (Fauré et al. 2006).

The use of nanomaterials in the water purification do not have a direct impact on humans, but these nano particles may find an access to human body through the consumption of sea food. Therefore the effects of nanomaterials on the aquatic life should be taken into account as it is an important question that is concerned with the toxicity of NPs. Nanoparticles including silver, titanium and carbon nanotubes are mainly causing adverse effects on aquatic life (Asghari et al. 2012; Clemente et al. 2012; Jackson et al. 2013).

The US Environmental Protection Agency in 2010 reported the effects of titanium dioxide nanoparticles on aquatic organisms. In this study, a comparison was made among/between the different nanoscale titanium dioxide, point of entry to biological systems and their harmful effects on organisms. The study concluded several toxic effects on organisms including reduction in reproduction rate, respiratory distress, pathological and behavioral changes (Jackson et al. 2013).

9.6 Conclusion

At present the foremost challenges which are faced by the world are the shortage of portable water and water pollution, resulting from a great raise in population and over exploitation of fresh water etc. To overcome these challenges at global level, there is an urgent need to introduce some advanced and practically achievable water decontamination techniques to ensure the high quality drinking water. Recent innovations in nanotechnology have provided highly efficient, advanced and cost effective means for the treatment and purification of water like nano-adsorption, nano-filtration etc. The development and use of nano materials for water remediation to improvise the water quality for better living has proved to be a main area of interest that is gaining importance throughout the world. The unique characteristic feature of nanomaterials in combination with convention methods enabled the decontamination process much reasonable and conventional and seems to be a most promising approach for future. Though the use of nanotechnology in the treatment of water has provided many advantages over the traditional methods, on the same time there are some challenges including nano-toxicity which limits their use in the water treatment. The effects of nano engineered materials on the living organisms need to be addressed to ensure the safety of aquatic life as well as human beings. In order to minimize the adverse impacts of nanomaterials on biological systems their toxicity must be assessed prior their release in the environment.

References

Ahmad A, Majid M, Ooi B (2011) Functionalized PSf/SiO_2 nanocomposite membrane for oil-in-water emulsion separation. Desalination 268(1–3):266–269

Akin I, Arslan G, Tor A, Ersoz M, Cengeloglu Y (2012) Arsenic (V) removal from underground water by magnetic nanoparticles synthesized from waste red mud. J Hazard Mater 235:62–68

Ali I, Jain C (1998) Groundwater contamination and health hazards by some of the most commonly used pesticides. Curr Sci 75(10):1011–1014

Aredes S, Klein B, Pawlik M (2012) The removal of arsenic from water using natural iron oxide minerals. J Clean Prod 29:208–213

Arthanareeswaran G, Devi TS, Raajenthiren M (2008) Effect of silica particles on cellulose acetate blend ultrafiltration membranes: Part I. Sep Purif Technol 64(1):38–47

Asghari S, Johari SA, Lee JH, Kim YS, Jeon YB, Choi HJ, Moon MC, Yu IJ (2012) Toxicity of various silver nanoparticles compared to silver ions in *Daphnia magna*. J Nanobiotech 10(1):14

Azrague K, Aimar P, Benoit-Marquie F, Maurette M (2007) A new combination of a membrane and a photocatalytic reactor for the depollution of turbid water. Appl Catal Environ 72(3–4):197–204

Bae TH, Tak TM (2005) Effect of TiO_2 nanoparticles on fouling mitigation of ultrafiltration membranes for activated sludge filtration. J Membr Sci 249(1–2):1–8

Banerjee SS, Chen DH (2007) Fast removal of copper ions by gum arabic modified magnetic nano-adsorbent. J Hazard Mater 147(3):792–799

Brammer H, Ravenscroft P (2009) Arsenic in groundwater: a threat to sustainable agriculture in South and South-east Asia. Environ Int 35(3):647–654

Brittany L, Carino V, Kuo J, Leong L, Ganesh R (2006) Adsorption of organic Compounds to metal oxide nanoparticles. Paper presented at the Conference presentation is part of: General Environmental

Cao X, Ma J, Shi X, Ren Z (2006) Effect of TiO_2 nanoparticle size on the performance of PVDF membrane. Appl Surf Sci 253(4):2003–2010

Chang YC, Chen DH (2005) Adsorption kinetics and thermodynamics of acid dyes on a carboxymethylated chitosan-conjugated magnetic nano-adsorbent. Macromol Biosci 5(3):254–261

Chen CY, Chang JC, Chen AH (2011) Competitive biosorption of azo dyes from aqueous solution on the templated crosslinked-chitosan nanoparticles. J Hazard Mater 185(1):430–441

Chorawalaa KK, Mehta MJ (2015) Applications of nanotechnology in wastewater treatment. Int J Innov Emerg Res Eng 2(1):21–26

Chou WL, Yu DG, Yang MC (2005) The preparation and characterization of silver-loading cellulose acetate hollow fiber membrane for water treatment. Polym Adv Tech 16(8):600–607

Clemente Z, Castro V, Jonsson C, Fraceto L (2012) Ecotoxicology of nano-TiO_2–an evaluation of its toxicity to organisms of aquatic ecosystems. Int J Environ Res 6(1):33–50

Condon JB (2006) Surface area and porosity determinations by physisorption: measurements and theory. Elsevier, Amsterdam

De Volder MF, Tawfick SH, Baughman RH, Hart AJ (2013) Carbon nanotubes: present and future commercial applications. Science 339(6119):535–539

Deliyanni E, Bakoyannakis D, Zouboulis A, Matis K (2003) Sorption of As (V) ions by akaganeite-type nanocrystals. Chemosphere 50(1):155–163

Diallo MS, Christie S, Swaminathan P, Johnson JH, Goddard WA (2005) Dendrimer enhanced ultrafiltration. 1. Recovery of Cu (II) from aqueous solutions using PAMAM dendrimers with ethylene diamine core and terminal NH_2 groups. Environ Sci Tech 39(5):1366–1377

El Rahman A, Gepreel M (2013) Nanotechnology applications in water treatment: future avenues and challenges: a review. Paper presented at the 6th International Perspective on Water Resources and the Environment Confernce, Izmir, Turkey

EPA (1999) Alternative disinfectants and oxidants guidance manual. Disinfectant use in water treatment. EPA, Washington, DC

EPA (2009) Fact sheet. Emerging contaminants–nanomaterials. Solid waste and emergency response (5106P): EPA 505-F-09-011. United States Environmental Protection Agency, Washington, DC

Fauré J, Lachenal G, Hirrlinger J, Chatellard-Causse C, Blot B, Grange J, Schoehn G, Goldberg Y, Boyer V, Kirchhoff F, Raposo G, Garin J, Sadoul R (2006) Exosomes are released by cultured cortical neurones. Mol Cell Neurosci 31(4):642–648

Faust SD, Aly OM (2018) Chemistry of water treatment. CRC Press, Boca Raton, FL

Feng C, Khulbe K, Matsuura T, Tabe S, Ismail A (2013) Preparation and characterization of electro-spun nanofiber membranes and their possible applications in water treatment. Sep Purif Technol 102:118–135

Fernandes SC, Sadocco P, Alonso-Varona A, Palomares T, Eceiza A, Silvestre AJ, Mondragon I, Freire CS (2013) Bioinspired antimicrobial and biocompatible bacterial cellulose membranes obtained by surface functionalization with aminoalkyl groups. ACS Appl Mater Interfaces 5(8):3290–3297

Friedmann D, Mendive C, Bahnemann D (2010) TiO_2 for water treatment: parameters affecting the kinetics and mechanisms of photocatalysis. Appl Catal Environ 99(3–4):398–406

Gao Y, Wahi R, Kan A, Falkner J, Colvin V, Tomson M (2004) Adsorption of cadmium on anatase nanoparticles effect of crystal size and pH. Langmuir 20(22):9585–9593

Gehrke I, Geiser A, Somborn-Schulz A (2015) Innovations in nanotechnology for water treatment. Nanotech Sci Appl 8:1

Goodman GL (2011) Gilman's the pharmacological basis of therapeutics. McGraw-Hill, New York, NY

Goswami A, Raul P, Purkait M (2012) Arsenic adsorption using copper (II) oxide nanoparticles. Chem Eng Res Design 90(9):1387–1396

Hashim DP, Narayanan NT, Romo-Herrera JM, Cullen DA, Hahm MG, Lezzi P, Suttle JR, Kelkhoff D, Muñoz-Sandoval E, Ganguli S, Roy AK, Smith D, Vajtai R, Sumpter BG, Meunier V, Terrones H, Terrones M, Ajayan PM (2012) Covalently bonded three-dimensional carbon nanotube solids via boron induced nanojunctions. Sci Rep 2:363

Hoet PH, Brüske-Hohlfeld I, Salata OV (2004) Nanoparticles–known and unknown health risks. J Nanobiotech 2(1):12

Hossain F, Perales-Perez OJ, Hwang S, Roman F (2014) Antimicrobial nanomaterials as water disinfectant: applications, limitations and future perspectives. Sci Total Environ 466–467:1047–1059

Jackson P, Jacobsen NR, Baun A, Birkedal R, Kühnel D, Jensen KA, Vogel U, Wallin H (2013) Bioaccumulation and ecotoxicity of carbon nanotubes. Chem Cent J 7(1):154

Jagadevan S, Jayamurthy M, Dobson P, Thompson IP (2012) A novel hybrid nano zerovalent iron initiated oxidation–biological degradation approach for remediation of recalcitrant waste metalworking fluids. Water Res 46(7):2395–2404

Jain P, Pradeep T (2005) Potential of silver nanoparticle-coated polyurethane foam as an antibacterial water filter. Biotech Bioeng 90(1):59–63

Jung JY, Chung YC, Shin HS, Son DH (2004) Enhanced ammonia nitrogen removal using consistent biological regeneration and ammonium exchange of zeolite in modified SBR process. Water Res 38(2):347–354

Karim MR, Rhodes ER, Brinkman N, Wymer L, Fout GS (2009) New electropositive filter for concentrating enteroviruses and noroviruses from large volumes of water. Appl Environ Microbiol 75(8):2393–2399

Keuter V, Gehrke I (2012) Development of multibarrier systems consisting of nano-enhanced membranes and UV-leds for water purification applications. Proc Eng 44:544–545

Khajeh M, Laurent S, Dastafkan K (2013) Nanoadsorbents: classification, preparation, and applications (with emphasis on aqueous media). Chem Rev 113(10):7728–7768

Kim ES, Deng B (2011) Fabrication of polyamide thin-film nano-composite (PA-TFN) membrane with hydrophilized ordered mesoporous carbon (H-OMC) for water purifications. J Membr Sci 375(1–2):46–54

Kim ES, Hwang G, El-Din MG, Liu Y (2012) Development of nanosilver and multi-walled carbon nanotubes thin-film nanocomposite membrane for enhanced water treatment. J Membr Sci 394:37–48

Krasner SW, Weinberg HS, Richardson SD, Pastor SJ, Chinn R, Sclimenti MJ, Onstad GD, Thruston AD (2006) Occurrence of a new generation of disinfection byproduct. Environ Sci Technol 40:7175–7185

Krishnan S, Weinman CJ, Ober CK (2008) Advances in polymers for anti-biofouling surfaces. J Mater Chem 18(29):3405–3413

Lalia BS, Guillen E, Arafat HA, Hashaikeh R (2014) Nanocrystalline cellulose reinforced PVDF-HFP membranes for membrane distillation application. Desalination 332(1):134–141

Langlet M, Permpoon S, Riassetto D, Berthomé G, Pernot E, Joud JC (2006) Photocatalytic activity and photo-induced superhydrophilicity of sol–gel derived TiO_2 films. J Photochem Photobiol A Chem 181(2–3):203–214

Lee J, Chae HR, Won YJ, Lee K, Lee CH, Lee HH, Kim I-C, Lee JM (2013) Graphene oxide nanoplatelets composite membrane with hydrophilic and antifouling properties for wastewater treatment. J Membr Sci 448:223–230

Lewinsky AA (2007) Hazardous materials and wastewater: treatment, removal and analysis. Nova Publishers, New York, NY

Li Q, Mahendra S, Lyon DY, Brunet L, Liga MV, Li D, Alvarez PJJ (2008) Antimicrobial nanomaterials for water disinfection and microbial control: potential applications and implications. Water Res 42:4591–4602

Li Z, Greden K, Alvarez PJ, Gregory KB, Lowry GV (2010) Adsorbed polymer and NOM limits adhesion and toxicity of nano scale zerovalent iron to *E. coli*. Environ Sci Tech 44(9):3462–3467

Liu X, Wang M, Zhang S, Pan B (2013) Application potential of carbon nanotubes in water treatment: a review. J Environ Sci 25(7):1263–1280

Masel RI, Masel RI (1996) Principles of adsorption and reaction on solid surfaces, vol 3. John Wiley and Sons, New York, NY

Matlochová A, Plachá D, Rapantová N (2013) The application of nanoscale materials in groundwater remediation. Pol J Environ Stud 22(5):1401

Mautner A, Lee KY, Lahtinen P, Hakalahti M, Tammelin T, Li K, Bismarck A (2014) Nanopapers for organic solvent nanofiltration. Chem Commun 50(43):5778–5781

Meng ZD, Zhu L, Ye S, Sun Q, Ullah K, Cho KY, Oh WC (2013) Fullerene modification CdSe/TiO_2 and modification of photocatalytic activity under visible light. Nanoscale Res Lett 8(1):189

Mohammed AS, Kapri A, Goel R (2011) Heavy metal pollution: source, impact, and remedies. In: Biomanagement of metal-contaminated soils. Springer, Dordrecht, pp 1–28

Nagy A, Harrison A, Sabbani S, Munson RS Jr, Dutta PK, Waldman WJ (2011) Silver nanoparticles embedded in zeolite membranes: release of silver ions and mechanism of antibacterial action. Int J Nanomedicine 6:1833

Nel A, Xia T, Mädler L, Li N (2006) Toxic potential of materials at the nanolevel. Science 311(5761):622–627

Nowack B, Krug HF, Height M (2011) 120 years of nanosilver history: implications for policy makers. ACS Publications, Washington, DC

Oberdörster G, Maynard A, Donaldson K, Castranova V, Fitzpatrick J, Ausman K, Carter J, Karn B, Kreyling W, Lai D, Olin S, Monteiro-Riviere N, Warheit D, Yang H, ILSI Research Foundation/Risk Science Institute Nanomaterial Toxicity Screening Working Group (2005) Principles for characterizing the potential human health effects from exposure to nanomaterials: elements of a screening strategy. Part Fibre Toxicol 2(1):1

Ostad E, Wise GJ (2005) Celestial bodies and urinary stones: Isaac Newton (1641–1727)–health and urological problems. BJU Int 95(1):24–26

Pacheco S, Tapia J, Medina M, Rodriguez R (2006) Cadmium ions adsorption in simulated wastewater using structured alumina–silica nanoparticles. J Noncrys Solids 352(52–54):5475–5481

Pan B, Xing B (2008) Adsorption mechanisms of organic chemicals on carbon nanotubes. Environ Sci Tech 42(24):9005–9013

Pandey J, Khare R (2016) Biomimetic nanotechnology: putting life into materials. Int J Adv Res Eng Appl Sci 5(8):44–53

Patil BBT (2015) Wastewater treatment using nanoparticles. J Adv Chem Eng 5:1–2

Pendergast MM, Dorin RM, Phillip WA, Wiesner U, Hoek EM (2013) Understanding the structure and performance of self-assembled triblock terpolymer membranes. J Membr Sci 444:461–468

Perreault F, De Faria AF, Elimelech M (2015) Environmental applications of graphene-based nanomaterials. Chem Soc Rev 44(16):5861–5896

Petrik L, Missengue R, Fatoba O, Tuffin M, Sachs J (2012) Silver/zeolite nano composite-based clay filters for water disinfection. WRC report (KV 297/12). Water Research Commission, Gezina. ISSN 1790670070

Qiu X, Yu H, Karunakaran M, Pradeep N, Nunes SP, Peinemann KV (2012) Selective separation of similarly sized proteins with tunable nanoporous block copolymer membranes. ACS Nano 7(1):768–776

Qu X, Brame J, Li Q, Alvarez PJ (2012) Nanotechnology for a safe and sustainable water supply: enabling integrated water treatment and reuse. Acc Chem Res 46(3):834–843

Qu X, Alvarez PJ, Li Q (2013) Applications of nanotechnology in water and wastewater treatment. Water Res 47(12):3931–3946

Quang DV, Sarawade PB, Jeon SJ, Kim SH, Kim JK, Chai YG, Kim HT (2013) Effective water disinfection using silver nanoparticle containing silica beads. Appl Surf Sci 266:280–287

Rao GP, Lu C, Su F (2007) Sorption of divalent metal ions from aqueous solution by carbon nanotubes: a review. Sep Purif Technol 58(1):224–231

Ren X, Zhao C, Du S, Wang T, Luan Z, Wang J, Hou D (2010) Fabrication of asymmetric poly (m-phenylene isophthalamide) nanofiltration membrane for chromium (VI) removal. J Environ Sci 22(9):1335–1341

Sadeghi KM, Arami M, Gharanjig K (2013) Dye removal from colored-textile wastewater using chitosan-PPI dendrimer hybrid as a biopolymer: optimization, kinetic, and isotherm studies. J Appl Polym Sci 127(4):2607–2619

Samanta HS, Das R, Bhattachajee C (2016) Influence of nanoparticles for wastewater treatment-a short review. Austin Chem Eng 3(3):1036

Sandia National Laboratories (2003) Desalination and water purification roadmap–a report of the executive committee. DWPR program report. US Department of the Interior, Bureau of Reclamation and Sandia National Laboratories, Albuquerque

Savage N, Diallo MS (2005) Nanomaterials and water purification: opportunities and challenges. J Nanopart Res 7(4–5):331–342
Savić R, Luo L, Eisenberg A, Maysinger D (2003) Micellar nanocontainers distribute to defined cytoplasmic organelles. Science 300(5619):615–618
Shao-feng N, Yong L, Xin-hua X, Zhang-hua L (2005) Removal of hexavalent chromium from aqueous solution by iron nanoparticles. J Zhejiang Univ Sci B 6(10):1022–1027
Shen L, Zhao B, Zhang J, Chen J, Zheng H (2010) Virus adsorption onto nano-sized iron oxides as affected by different background solutions. Huan jing ke xue= Huanjing kexue 31(4):983–989
Singh R, Smitha MS, Singh SP (2014) The role of nanotechnology in combating multi-drug resistant bacteria. J Nanosci Nanotechnol 14(7):4745–4756
Skubal L, Meshkov N, Rajh T, Thurnauer M (2002) Cadmium removal from water using thiolactic acid-modified titanium dioxide nanoparticles. J Photochem Photobiol A Chem 148(1–3):393–397
Stoimenov PK, Klinger RL, Marchin GL, Klabunde KJ (2002) Metal oxide nanoparticles as bactericidal agents. Langmuir 18(17):6679–6686
Sun C, Qu R, Ji C, Wang C, Sun Y, Yue Z, Cheng G (2006) Preparation and adsorption properties of crosslinked polystyrene-supported low-generation diethanolamine-typed dendrimer for metal ions. Talanta 70(1):14–19
Tang C, Zhao Y, Wang R, Hélix-Nielsen C, Fane A (2013) Desalination by biomimetic aquaporin membranes: review of status and prospects. Desalination 308:34–40
Taurozzi JS, Arul H, Bosak VZ, Burban AF, Voice TC, Bruening ML, Tarabara VV (2008) Effect of filler incorporation route on the properties of polysulfone–silver nanocomposite membranes of different porosities. J Membr Sci 325(1):58–68
Tiwari DK, Behari J, Sen P (2008) Application of nanoparticles in waste water treatment 1. World Appl Sci J 3(3):417
Usmania MA, Khan I, Bhatd A, Pillaie RS, Ahmadf N, Haafizg MM, Ovesh M (2017) Current trend in the application of nanoparticles for waste water treatment and purification: a review. Curr Organ Synth 14(2):206–226
Vecitis CD, Zodrow KR, Kang S, Elimelech M (2010) Electronic-structure-dependent bacterial cytotoxicity of single-walled carbon nanotubes. ACS Nano 4(9):5471–5479
Wegmann M, Michen B, Graule T (2008) Nanostructured surface modification of microporous ceramics for efficient virus filtration. J Eur Ceram Soc 28(8):1603–1612
Whitesides GM, Grzybowski B (2002) Self-assembly at all scales. Science 295(5564):2418–2421
Xie W, He F, Wang B, Chung TS, Jeyaseelan K, Armugam A, Tong YW (2013) An aquaporin-based vesicle-embedded polymeric membrane for low energy water filtration. J Mater Chem A 1(26):7592–7600
Yang HY, Han ZJ, Yu SF, Pey KL, Ostrikov K, Karnik R (2013) Carbon nanotube membranes with ultrahigh specific adsorption capacity for water desalination and purification. Nat Commun 4:2220
Zhao S, Wang P, Wang C, Sun X, Zhang L (2012) Thermostable PPESK/TiO_2 nanocomposite ultrafiltration membrane for high temperature condensed water treatment. Desalination 299:35–43

Chapter 10
Applications of Nanobiosensors in Agriculture

Nadia Ghaffar, Muhammad Akhyar Farrukh, and Shagufta Naz

10.1 General Introduction

The backbone of most of the developing countries is agriculture. It's not only satisfying the hunger of community but also have a great impact on economy. Accessibility of water and land resources for agriculture is decreasing rapidly that leads to decline in agriculture productivity.

There are different factors which affect the agricultural output like deficiency of macro and micro nutrients, depletion of water resources, population explosion and difference in soil condition, industrialization and erosion of top soil. Soil lacks some macro and micro nutrients therefore fertilizers are used in agriculture to supplement soils with these nutrients. In developing countries 35–40% of the crop yield depends upon fertilizer but few fertilizers directly affects the growth of plants. In addition, the concentration of heavy metals, herbicides, insecticides, pesticides and fertilizers in agricultural land is on alarming stage (Corradini et al. 2010).

For example annually 4% growth in agriculture is earmarked by the Indian national policy makers to ensure national food security with the inadequate availability of water and land resources. Avenue for achieving the demand is possible only by enhancing the yield and income per unit of the limited supply of natural resources by using improved technologies efficiently (Chena and Yada 2011).

New technologies are needed in agriculture to enhance the productivity of crops. Nano agriculture is an area which involves the synthesis and implication of nanoparticles in agriculture. These synthesized particles play an important role in improving

N. Ghaffar (✉)
Department of Botany, Lahore College for Women University, Lahore, Pakistan

M. A. Farrukh
Department of Chemistry, Foramen College, The Chartered University, Lahore, Pakistan

S. Naz
Department of Biotechnology, Lahore College for Women University, Lahore, Pakistan

S. Javad (ed.), *Nanoagronomy*, https://doi.org/10.1007/978-3-030-41275-3_10

the crop productivity (Srilatha 2011). Nanotechnology deals with the smallest nanoparticles having a size of 1–100 nm, that has potential to enhance the agricultural yield by solving the problems that remain unsolved due to conventional methods. Nanotechnology and nanoscale science has potential to provide better and novel solution to all the challenges that agriculture is facing today (Manjunatha et al. 2016).

The development of Nanomaterials and Nanodevices provides different horizons for improving agricultural system. It has the potential to revolutionize the agricultural systems. These nanomaterials have a lot of novel applications in agriculture and biotechnology (Srilatha 2011).

Nanobiosensors can be used efficiently for sensing soil pH, moisture and a range of pathogens, insecticide, herbicide, pesticide and fertilizers. Appropriate use of nanobiosensor may help in supporting sustainable agriculture for improving crop productivity (Rai et al. 2012).

10.2 Biosensors

Biosensor is basically an integrated self-contained device which uses a biological recognition element and provides precise quantitative or semi quantitative analytical information. Biosensors are used to detect biological product by combining physical or chemical transducer with a biological recognition element. Researchers are working on integration of nanoparticles into materials that are used in construction of biosensors, so that sensitivity and performance of the system may improve. Advanced form of biosensors can play an important role in existing and potential sensing applications (Turner 2000). Some of the data is summarized in Table 10.1, showing evolutionary history of biosensors.

10.3 Nanobiosensors

Nanotechnology is a blistering field of science that deals with the production and modification of materials to a nanosize (10^{-9} m) range. On the basis of nanotechnology many advanced nanobiosensor with a highly miniature size have been developed in twenty-first century. Researchers are still working and recently they have generated biosensors by combining computer, nanaoscience, biology and electronics. These advanced biosensors are more reliable with high resolution depicting extraordinary sensing capabilities.

Nanobiosensors are basically modified form of biosensors with compact analytical unit which contains biologically derived sensitizing element and a physiochemical transducer attached with it. Invention of first biosensor was done in 1967 which lead to the invention of new and modified forms of biosensors. There are basically

Table 10.1 Milestone in the advancement of biosensors in twentieth century (Rai et al. 2012)

Year	Advancement of biosensors
1906	Data about protein Immobilization was reported
1922	First pH electrode made up of glass was prepared
1956	Information about first oxygen electrode was arranged
1962	First time biosensor was described
1969	First potensiometric biosensor was seen
1970	Ion selective field effect transistor (ISFET) was invented
1974	Reports on thermal transducers for biosensor
1975	To describe fiber optic sensor the term optode was invented
1976	Electrochemical glucose biosensor incorporation in bedside artificial pancreas
1977	The term biosensor was invented by Karl Camman
1982	Illustration of first needle type enzyme electrode
1984	First paper was published on the use of ferrocene and its derivatives as an immobilized mediator for use with oxidoruductases
1987	For home blood glucose monitoring a screen printed enzyme electrode was initiated by MediSense
1990	Pharmacia BIA Core SPR based biosensor system was instigated
1992	Hand held blood analyzer launched by i-STAT
1996	Invention of Glucocard
1998	Blood glucose biosensor was launched
2002	Use of electro deposition paints for biosensors as immobilization matrices
2003	In living plant an enzymatic glucose/O_2 fuel cell was entrenched
2007	First time implanted glucose biosensor was operated for 5 days

three generations of biosensors which resulted in development of several modern and innovative biosensors (Updike and Hicks 1967).

Nanosensors are combined with immobilized bioreceptor probes and are much selective for targeted molecules. Nanobiosensors are constructed in nanoscale range to evaluate the data at atomic scale level (Freitas 1999. http://www.nanome-dicine.com/NMI/Glossary.htm). In past it was not possible to examine the real bio-analytical applications but nanobiosensors provide new horizons for basic research and for bio-analytical applications. For molecular analysis nanobiosensors can be integrated into lab-on-a-chip, they not only assist in molecular analysis but also help in monitoring of many metabolites. They can detect a range of microorganisms, many analytes like insecticides, urea and glucose etc. Due to their ability to be easily carried they are perfect for field application and can also facilitate the laboratory settings (Fan et al. 2008; Khanna 2008; You et al. 2009; Velasco 2009).

10.3.1 *Why Nanobiosensors?*

When the size of particles is modified, their associated properties are also changed by increase or decrease in particle size. Decrease in particle size causes change in physics and chemistry of particles that leads to development of novel properties that totally depends upon size.

- Increase in surface to volume ratio is observed by decreasing the size.
- In order to make improved miniature structure device, size of the transducer and sensing part in a sensor must be reduced.
- New sensing devices are in the process of development that is totally inspired from nano materials science (Choudhary et al. 2015).

10.3.2 *Intrinsic Worth of Nanobiosensors over Conventional Biosensors*

Nanobiosensors are very sensitive and reliable as compared to conventional biosensors, they can sense a single virus particle, and even they can detect much less concentration of any harmful substance. The reasons for this efficient role may be because

- Nanobiosensors works at atomic scale and depicts much efficiency
- Nanobiosensors also have increased surface to volume ratio
- Nanobiosensors are extremely sensitive with very low chances of error.
- But nanobiosensors are still under developmental stage and need an elaborative research to find more applications in the field (Rai et al. 2012).

10.3.3 *Distinctive Properties of Nanobiosensors*

Rai et al. depicted some unique properties of nanobiosensors which are given below in Fig. 10.1. These are some model requirements for nanobiosensors.

10.4 Components of Best Nanobiosensors

It contains three important parts (Fig. 10.2)

1. The sensitive biological element (e.g. cell receptors, nucleic acids, tissue, antibodies, microorganisms, enzymes, cell receptors etc.), that combine with or

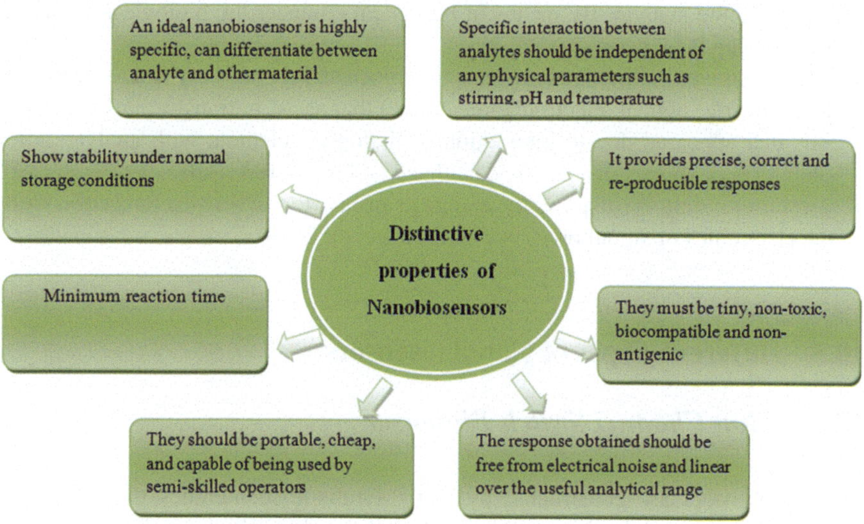

Fig. 10.1 Distinctive properties of biosensors

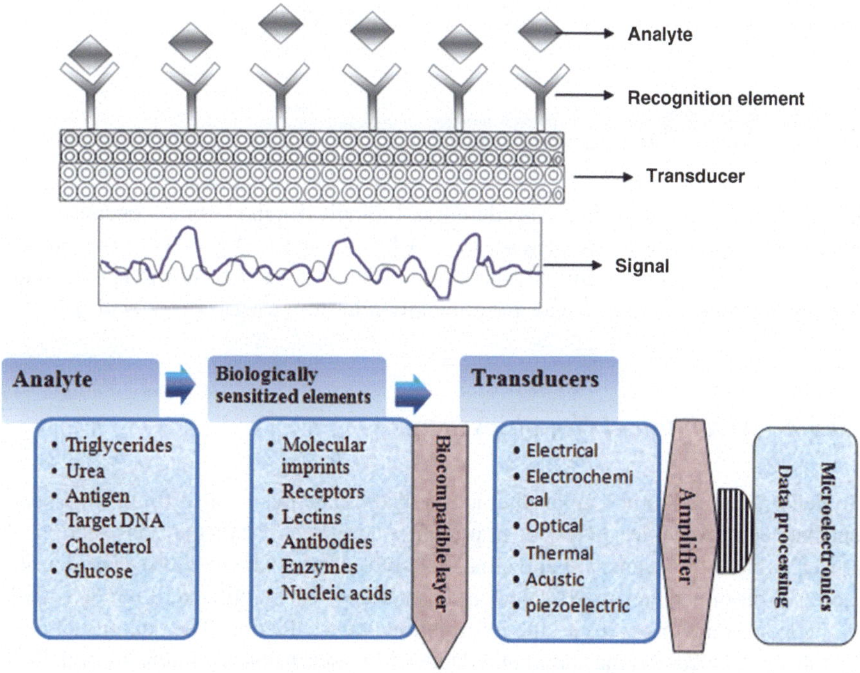

Fig. 10.2 (**a**) Schematic presentation of a biosensor. (**b**) Principle of nanobiosensor

recognize the analyte. By means of biological engineering, biological sensitive element can be generated.
2. The transducer or the detector element (mechanism may be physicochemical; piezoelectric, optical, electrochemical, etc.) that assist in transformation of signals that appeared due to interaction of biological element with analytes into such form of signals that can be calculated and measured easily.
3. Biosensor monitors (displays the results in an easy to use way in connection with the electronics or signal processors) (Ejeian et al. 2018).

10.5 Different Types of Nanobiosensors

10.5.1 Ion Channel Switch Biosensor

These biosensors are constructed by using synthetic membranes that are self-assembled, assist in detection of signals, and behaves like a biological switch. They can detect the presence of a particular molecule by generating electrical current. They are very specific, provides particular and quantitative information immediately. They reduce the delivery time of test results from hours to minuets and play an important role in emergency diagnoses (Cornell 2002).

10.5.2 Viral Nanobiosensors

Basically virus particles are considered as biological nanoparticles because they have some applications as nanosensors. Adenovirus and HSV (Herpes simplex virus) are two of many examples that are used to activate the assemblage of magnetic nanobeads as a nanosensor for clinically relevant viruses (Perez et al. 2003).

10.5.3 Mechanical Nanobiosensors

To determine the biomolecular interaction, Nanoscale mechanical forces that present between molecules plays an important role and assist in the construction of advance, label free, sensitive and small size biosensors (Cheng 2006). For identification of biomolecules microscale cantilever beams can be utilized, it interacts with biomolecules and provides all the information after deflection. The quantity of analyte in solution can be measured and deduced by measuring the intensity of deflection of cantilever beams. There are three methods that help in conversion of concerned analyte into micro mechanical bending of cantilever, (a) surface stress may cause bending, (b) bending due to mass loading, (c) change in temperature may cause bending. All the nano mechanical appliances are extremely mass sensitive. Reduction in size causes reduction in mass, consequently a prominent change in main mass is noticed after adding bounded analyte molecule (Ziegler 2004).

10.5.4 Optical Nanobiosensors

Construction of optical biosensors depends upon arrangement of optics where light rays dispersed and circulate in a closed path. When analyte connects to resonator, a change is observed which is documented in resonant frequency. There are basically two types of resonators, (1) linear resonator and (2) ring resonator. Linear resonator contains two end mirrors; light rays hit the mirrors and reflect back while in ring resonators end mirrors are not present so light dispersed in two different directions.

In nanobiosensor optical fibers of small size are used that plays an important role in sensing. They help in intercellular, intracellular sensing and also allow physiological and biological parameters (Erickson et al. 2008).

10.5.5 Electronic Nanobiosensors

These nanobiosensors are based on microchips. Every chip contains many sensors which can work independently. Their function is to detect the target DNA and its binding that actually create a bridge on a microchip between two electrically separated wires. These nanobiosensors can be separately deal with captured probes for targeted DNA molecules that comes from different or from same organism (Jain 2005).

10.5.6 PEBBLE Nanobiosensors

Probes Encapsulated by Biologically Localized Embedding (PEBBLE) nanobiosensors contains many sensors that are basically produced by micro emulsion polymerization method. It produces sensors of spherical shape which ranges from 20 to 200 nm and is enclosed in a chemically inert matrix. Many different types of sensor molecules can be trapped in the matrix that can assist in detection of fluorescence, can detect change in pH, can detect optical change and change in Ca^{2+} ions (Sumner et al. 2002). These nanobiosensors are very proficient and can examine real-time intracellular and intercellular imaging of molecules and ions. They are not sensitive to interference from proteins so it can move back to its base line and also demonstrate constancy to photo bleaching. They exhibit a strong oxygen sensing capacity in human plasma and are very less affected by scattering of light and auto fluorescence (Cao et al. 2004).

10.5.7 Nanoshell Biosensors

Nanoshell biosensors are very efficient and their working is based on detection of analyte from complex biological media. For detection of analyte, no sample preparation is needed; simply gold nanoshells are placed in a rapid immunoassay. Nanoshell/antibody aggregates forms conjugation with extinction spectra that can be examined in the presence of analyte with the help of spectroscopy. Nanoshells are very proficient and can improve the chemical sensing to a great extent as ten billion times (Choudhary et al. 2015).

10.5.8 Nanowire Biosensors

These biosensors are basically comprised of two extremely sensitive molecules: carbon nanotube and a single stranded DNA molecule. Carbon nanotube works as a transmitter and single stranded DNA works like a detector. Nanowires can behave independent of analyte if we modify their surface properties. These properties can easily be modified by utilizing biological or chemical ligands. By changing the surface properties chemical binding started on their surface that leads to change in conductance of the nanowire in a sensitive manner. Highly sensitive and real time electrical sensors can be produced by using Boron-doped silicon nanowires (SiNWs) for biological species (Choudhary et al. 2015).

10.6 Applications of Nanobiosensor in Agriculture

Currently, nanobiosensors are presenting attractive prospects in comparison to traditional biosensors. Nanobiosensors have some distinctive properties and advantages. They are more sensitive, proficient, and specific and have enhanced detection system. They have some potential applications in different fields including agriculture, bioprocess control, bio defense, quality control of food and the most important medical applications. Here we are concerned about potential applications of nanobiosensors in agriculture and agro products. Few applications of nanobiosensors are explained below in Fig. 10.3.

10.6.1 Evaluation of Disease and Soil Quality

Nanosensors can be utilized to identify the soil diseases that are caused by microorganisms present in soil like bacteria, viruses and fungi. Nanosensors diagnose the soil disease by quantitatively measuring the oxygen consumption of "good and bad

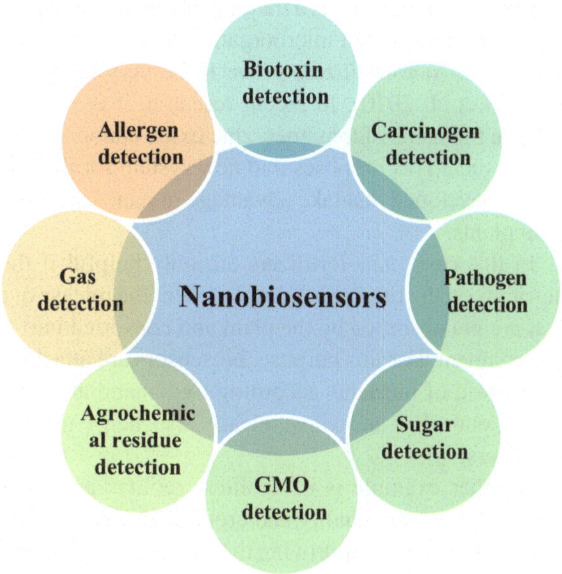

Fig. 10.3 Applications of nanobiosensor in agriculture

microbes" during their respiration in soil. The measurement completed by going through following steps:

1. In the first step two sensors will be selected and saturated with "good microbes" and "bad microbes" and then drenched in soil suspension in buffer solution.
2. In the second step both microbes will be examined for oxygen consumption and data will be recorded.
3. In the third step data collected from both microbes will be compared which will depict that which microbe is best for the soil (AIST 2004, http://www.aist.go.jp/aiste/latestresearch/2004/20040402_1/20040402_1.html)

10.6.2 Role of Biosensors to Support Sustainable Agriculture

A Nano fertilizer plays an important role in supporting the sustainable agriculture. A nano fertilizer is basically a product that is encapsulated within nanoparticles and provides nutrients to soil and crops. Encapsulated nano fertilizer can be prepared by three ways:

1. Encapsulation of nutrients can be done in the form of nanotubes or nano-porous materials.
2. Encapsulation can be done by coating the nutrients with a thin protective polymer film.
3. Nutrients can be isolated and dispersed as nanoscale emulsions or as particles.

Nitrogen deficiency is a major problem in some areas which happen due to leaching, absorption of soil microorganisms for longer period of time and may be due to emissions. Nanofertilizers prove to be very helpful in reducing the nitrogen loss (Derosa et al. 2010). In recent research, it is reported that carbon nanotubes can enter in the seeds of tomatoes, zinc oxide nanoparticles can penetrate in the ryegrass root tissues. This proposes that new system for the release and delivery of nutrients can be developed that take advantage of nanoscale porous domains on the surface of crop plants.

In this case, nanofertilizers are only helpful if they can control the release of nutrients on the time of need and prevent the release of nutrients without any demand that are not absorbed by the plant and converted into gaseous form and polluted the environment. For this purpose, biosensors are attached to nanofertilizers that permit the release of nutrients according to demand and according to soil nutrient condition. Controlled and slow release also helps in reducing the harmful effects of fertilizers.

Another example is of zeolites, zeolites are thought to be very supportive in enhancing the soil quality and crop yield. Zeolites are basically aluminum silicates crystals that help in improving the soil quality, plant growth and crop yield, improves the water retention, infiltration and efficiency of fertilizer, retain nutrients in the soil for longer period of time for the plant and lessen the nutrient loss from soil.

Zeolite is considered to retain the nutrients in the roots of plants and supply to plant when it is needed. It assists the plant for proficient use of K and N fertilizers and fertilizer will not be wasted and even small amount of fertilizer can produce a better yield. Zeolite remain in the soil for longer period of time and assist in making better nutrients quality and retention time.

A combination of Zeolite and nanobiosensor modernized the agriculture and plays a significant role in advancement of agriculture. Biosensor sense the deficiency of nutrient or water in the soil or plant and control their release that zeolite have retained.

Pesticides encapsulated in nanoparticles have been synthesized which release pesticides after a specific time and due to change in environment. Herbicides have also combined with nanobiosensors that are provided to plants only when needed that leads to better crop yield (Lin and Xing 2008).

10.6.3 For Storage of Seeds

When we store the seeds they release some volatile aldehydes due to ageing process. These volatile compounds are even deleterious for other seeds. By detecting these volatile aldehydes in seed storage area, deteriorated seeds separated from other seeds and refreshed before their use. For this purpose Electronic nose (E-Nose) is used (Choudhary et al. 2015). In the field of sensor technology Electronic Nose is the most advance device (Fig. 10.4). It can detect and classify the vapors, gases and odors automatically. Electronic Nose is a unique device in comparison to other ana-

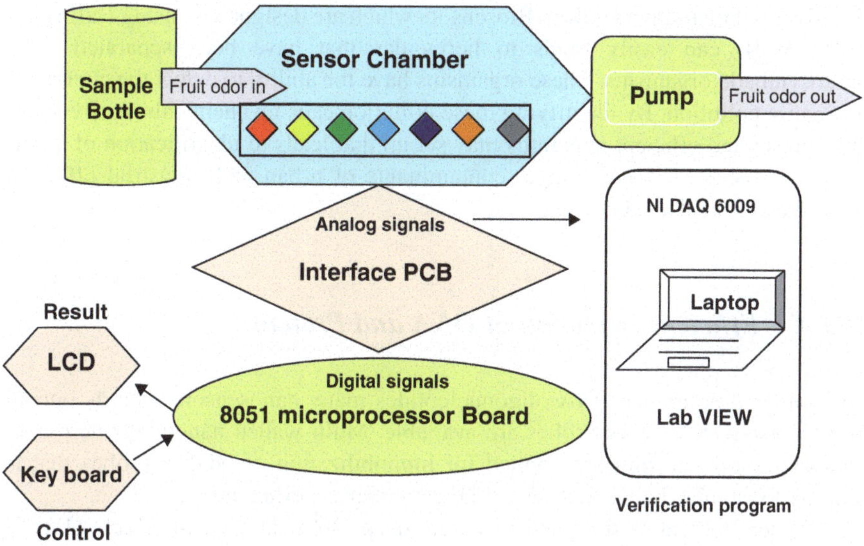

Fig. 10.4 Electronic nose system

lytical devices because of its exceptional detection system. It not only detects particular chemical specie but also recognize the fusion of organic sample (Kalita et al. 2015).

10.6.4 Smart Delivery Systems

Nano based smart delivery systems and nanobiosensors have a significant role in the utilization of natural resources like soil nutrients, water and other organic compounds. Farm managers can detect the crop pests and identify the stress (for example drought) by satellite photographing of fields through global positioning systems and nanomaterials. If crop pests and drought are identified, pesticide applications and irrigation levels will be modified automatically. Encapsulated fertilizers have capabilities to release the fertilizer slowly and according to need, that leads to minimize the fertilizer use and environmental pollution (Choudhary et al. 2015).

10.6.5 Detection of Contaminants and Other Molecules

Many nanobiosensors are constructed to identify the temperature, pressure, soil nutrients, stress, pests and other contaminants. Biosensors based on liposome are used to identify the Organophosphorus pesticides (paraoxon and dichlorvos). Zhang et al. (2007) reported a technique to identify the *Escherichia coli (E. coli)* by utilizing advanced bismuth nanofilm. Biosensors which are designed by using PSII (photosystem II) can easily binds to herbicides that have been separated from photosynthetic organisms. These organisms have the ability to detect the chemicals that cause pollution. By identifying these pollution causing chemicals, easily available, cheap and efficient apparatus may set up that leads to identification of a particular herbicide, sewage sludge, contaminants of urban and industrial effluents (Giardi and Piletska 2006).

10.6.6 Efficient Detection of DNA and Protein

To identify a particular DNA oligonucleotides many nanosensors such as optical biosensors/ssDNA-CNTs probes are available. Multi walled nanotube/zinc oxide/chitosan composite film are applied for immobilization of ssDNA probes that is used to distinguish between different DNA sequences efficiently.

A nanobiosensor is designed to detect deep DNA damage in which MWNT (Multi Walled Nanotube) bionanocomposite layer in chitosan is placed on screen printed carbon electrode (Ma et al. 2008). Transistor based nanowire field effect biosensor was reported by Maki et al. (2008) which avoids the PCR amplification and bisulfite treatment but can detect the DNA methylation. Biosensors based on protein nanoparticles interactions have been constructed which are ultra-sensitive and can detect a particular protein molecule. Biosensors which can detect the protein and DNA play significant role in identification of plant pathogens, anomaly in plants due to deficiency of some minerals and can easily differentiate one plant species from others.

10.6.7 Role of Biosensor in Effective Investigation of Food Products

Analysis of food products by using biosensor is becoming a very essential factor in food industry. Some of its applications are given below;

Analysis of vitamins: CM5 sensor chips are used for the immobilization of vitamin and the SPR (surface plasmon resonance) biosensor identify the interactions of these vitamins with a particular binding protein.

Detection of antibiotics: Some banned antibiotics have been detected in honey, due to that reason biosensors are now in frequent use. Biosensors can identify the presence of a particular antibiotic, efficiently and in a short period of time.

Detection of food spoilage: Amperometric biosensor based on immobilization of enzyme diamine oxidase, detect the amount of histamine in tiger prawn. In the same way potentiometric biosensor utilize 23 CO selective electrodes for detection of iso-citrate that leads to detection of spoilage.

Detection of microbial contamination: Contamination due to microbes can be detected by Immuno bio sensors. These biosensors use the immobilized monoclone antibodies onto indium tin oxide (ITO) electrodes, it can easily detect the presence of *Escherichia coli* O157:H7 (Rai et al. 2012).

10.6.8 Role of Biosensor in Detection of Plant Pathogen

Maurer et al. (2012) used a multi-arrayed sensor for plant pathogen detection as illustrated in Fig. 10.5.

10.6.9 For Detection of Herbicides

Nanobiosensors have been designed for the detection of herbicides by using atomic force microscopy tips with immobilized biomolecules that can combine with analyte. By measuring the force between analyte and biomolecule detection can be made.

Herbicides can be identified at a very less concentration by using nanobiosensors which are chemically reformed with antibodies and enzymes. Nanobiosensors are very sensitive and selective and can detect the presence of herbicides if a particular biomolecule interact with the target herbicide (Choudhary et al. 2015).

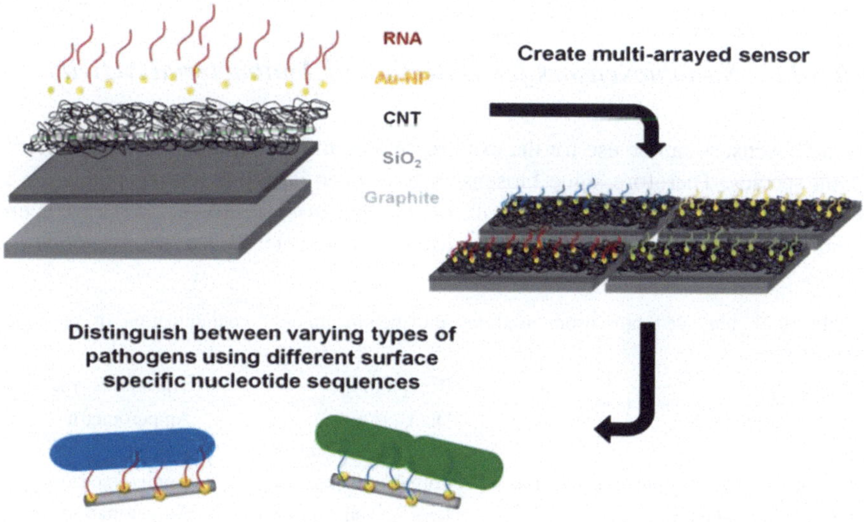

Fig. 10.5 Detection of plant pathogen (Maurer et al. 2012)

10.6.10 For Identification and Measurement of a Range of Toxics

To examine the soil and water toxicity generally microorganisms are used (like ToxAlert kit of Merck co.). For this purpose detection system basically depends upon luminescent bacteria, for example *vibrio fischeri* is used for the detection of toxic effects in biological test samples (Krystofova et al. 2010). Some biosensors have been designed for the detection of contamination in phenol containing compounds. Few phenolic compounds are given in Table 10.2.

10.6.11 Nanobiosensors as a New Tool for Herbal Pharmacology

Herbal pharmacology basically uses natural herbs and medicinal plants to support a better life. Now a day's development and advancement of all the crude extracts comes under pharmacology. Research on DNA-molecular markers is progressing in the world that leads to identification and investigation of some important herbal drugs, so that they may accept and can be used for medicinal purpose. Study on DNA analysis is thought to be very significant for plant pharmacology but it's not being used in the farm and is only limited to laboratory. For that reason nanoaptasensor have been designed, and it can behave as a good proposal for pharmacological applications (Secil 2008).

10.6.12 Nanobiosensors for Detection of Antibacterial Activity

Nanobiosensors can be use for the examination of antibacterial activity of different plant species. Therefore some biosensors have been designed and 16 plants have been studied by using ethanol as a solvent. Two bacterial strains (*E. coli*) have been modified from *E. coli* RFM 443 were used for antibacterial activity. Plasmids of

Table 10.2 Fabricated biosensors used for the identification of contamination of phenolic compounds

Analyte type	Receptor type	Transducer type
m-Cresol or catechol	DNA	Amperometric
Phenol	Mushroom tissues	Amperometric
Phenol, p-Cresol, m-Cresol or catechol	Polyphenol oxidase	Amperometric
Phenol/chlorophenol Catechol/phenol Cresol/chlorocresol Phenol/cresol	Laccase and tyrosinase	Amperometric Multicanal

both the modified *E. coli* strains contain luxoperon (structural genes) from *Vibrio fischeri* bacteria. Luminescence biosensors are basically applied for measuring antibacterial activity of all plants (Simona et al. 2011).

10.6.13 Nanobiosensors for Detection of Specific Secondary Metabolites

Plant secondary metabolites can be synthesized by using chemical method but this process is very expensive and sometimes it becomes very difficult and the worth of natural products can't be denied in comparison to synthetic products. So the plants with enough quantity of secondary metabolites are the best possible solution. In medicinal plants, quantity and quality of secondary metabolites is very important. There are different methods for the identification of secondary metabolites like GCMS (Gas chromatography mass spectrometry), HPLC (High Performance Liquid Chromatography) and GC (Gas chromatography) etc. but these techniques are not economic and are time consuming. Hence, the use of nanobiosensor is suggested because nanobiosensors are efficient and sensitive as compared to traditional system (Nabiabad et al. 2015).

10.6.14 Determination of Tropane Alkaloids

Tropane alkaloids can be identified by using biosensors, for that purpose optical biosensors (based on fluorescence) can be utilized. Biosensors based on FRET (Fluorescence Resonance Energy Transfer) system in which fluorescence probe consists of rodamin123 diflorescence and quantum dot's semi conductive nano crystals are basically applied for the identification of tropane alkaloids. The bio receptor used for this purpose is muscarinic M_2R receptor.

Baghery in 2013 described a nanobiosensor for the first time that can detect the presence of anti-cholinergic compounds in plant extracts. In another research it is reported that in plants monocrotophos (organophosphors) can be identified by using acetyl choline transferase enzyme-voltammetry biosensor (based on combination of gold nano particle-CdTe). For detection of glycol alkaloid, potentiostate biosensor that based on field-effect transistor and choline esterase can be applied (Baghery 2013).

10.6.15 Revealing the Fraud and Replacement in Herbal Drugs

Medicinal plants and herbal drugs have much importance in medical and in other fields. Use of some other plants and drugs has been noticed in place of original medicinal plants and herbal drugs. It is noticed that original plant species have been replaced by their close relative species that looks like almost same but their pharmacological properties are different. Due to this reason molecular markers (RFLP, RAPD and RFLP) have been introduced for identification and differentiation of original herbs from their close relatives. It is exposed by DNA fingerprinting that six species of composite plants were used in place of *Taraxacum mongolicum*. Besides confirmation of medicinal plants, it was also necessary to identify the quantity of phytochemicals so that they may use for drug purpose. Invention of DNA markers and DNA finger prints have much importance in detection of quantity and quality of active plant compounds that leads to quality control use in herbal drugs (Secil 2008). From all the above facts it is supposed that DNA sequences can be used as aptamers and apta biosensor can be constructed for identification and differentiation of photochemical more efficiently instead of molecular markers.

10.7 Future Perspectives

Nanotechnology is a blistering field and has potential to improve the environment, economy, agriculture and even medical fields. New prospects are under investigation for integration of nanotechnologies into nano-biosensors. With the combined efforts of researchers and government such facilitated agro products will flourish soon that leads to development of novel products. It is accepted that nanotechnology has the potential to transform the agriculture system. With the focused research and targeted efforts nanotechnology will take all the ways towards development of sustainable agriculture.

10.8 Conclusion

Nanobiosensors have great sensitivity. They are simple to use and needs less sample material. They are portable and have faster detection rates. Nanobiosensors increased the surface to volume ratio and are capable of measuring more variables. They consist of multi analyte and works at atomic scale with highest efficiency. But nanobiosensors are still under infancy stage and lacks nanoscale quantification. Nanobiosensors are needed for advanced and efficient agricultural system with economical use of resources, accurate detection of stress and good post harvest care.

References

AIST (2004) Latest research. http://www.aist.go.jp/aiste/latestresearch/2004/20040402_1/20040402_1.html

Baghery F (2013) Application of optical biosensor for measurement of tropan group alkaloids in the transgenic hairy root extraction of *Belladonna* plant. Dissertation, Bu Ali Sina University, Hamadan

Cao Y, Lee KYE, Kopelman R (2004) Polydecyl methacrylate based fluorescent PEBBLE swarm nanosensors for measuring dissolved oxygen in biosamples. Analyst 129:45–50

Chena H, Yada R (2011) Nanotechnologies in agriculture: new tools for sustainable development. Trends Food Sci Technol 22:585–594

Cheng MM (2006) Nanotechnologies for biomolecular detection and medical diagnostics. Curr Opin Chem Biol 19:10–11

Choudhary MK, Manvendra S, Vinod S (2015) Application of nanobiosensors in agriculture. Popular Kheti 3:130–135

Cornell BA (2002) Optical biosensors: present and future. Elsevier, Amsterdam, p 12

Corradini RE, Moura MR, Mattoso LHC (2010) A preliminary study of the incorporation of NPK fertilizer into chitosan nanoparticles express. Polym Lett 4:509–515

Derosa MC, Monreal C, Schnitzer M, Walsh R, Sultan Y (2010) Nanotechnology in fertilizers. Nat Nanotech 5:91

Ejeian F, Parisa E, Hajar AMT, Asieh S, Low ZX, Mohsen A, Asghar TK, Amir R (2018) Biosensors for waste water monitoring: a review. Biosens Bioelectron 118:66–79

Erickson D, Mandal S, Yang AHJ, Cordovez B (2008) Nanobiosensors: optofluidic, electrical and mechanical approaches to biomolecular detection at the nanoscale. Microfluid Nanofluid 4:33–52

Fan X, White IM, Shopova SI, Zhu H, Suter JD, Sun Y (2008) Sensitive optical biosensors for unlabeled targets: a review. Anal Chim Acta 620:8–26

Freitas RA Jr (1999) Nanomedicine. Volume I: Basic capabilities. Landes Bioscience, Georgetown, TX. http://www.nanomedicine.com/NMI/Glossary.htm

Giardi MT, Piletska EV (2006) Biotechnological applications of photosynthetic proteins: biochips, biosensors and biodevices. Biotechnology intelligence unit. Springer, New York, NY

Jain KK (2005) Nanotechnology in clinical laboratory diagnostics. Clin Chim Acta 358:37–54

Kalita P, Manash PS, Singh NH (2015) Electronic nose technology and its application. A systematic survey. Int J Innov Res Elect Electron Instrument Control Eng 3:123–128

Khanna VK (2008) New generation nano engineered biosensors, enabling nanotechnologies and nanomaterials. Sens Rev 28:39–45

Krystofova O, Trnkova L, Adam V, Zehnalek J, Hubalek J, Babula P, Kizek R (2010) Electrochemical microsensors for the detection of cadmium(II) and lead(II) ions in plants. J Biosens 1:5308–5328

Lin DH, Xing BS (2008) Root uptake and phytotoxicity of ZnO nanoparticles. Environ Sci Tech 42:5580–5585

Ma Y, Jiao K, Yang T, Sun D (2008) Sensitive PAT gene sequence detection by nano-SiO2/Paminothio phenol self-assembled films DNA electrochemical biosensor based on impedance measurement. Sens Actuat 131:565–571

Maki WC, Mishra NN, Cameron EG, Filanoski B, Rastogi SK, Maki GK (2008) Nano wiretransistor based ultra-sensitive DNA methylation detection. Biosens Bioelectron 23:780–787

Manjunatha SB, Biradarand DP, Aladaka YR (2016) Nanotechnology and its applications in agriculture: a review. Farm Sci 29:1–13

Maurer EI, Comfort KK, Hussain SM, Schlager JJ, Mukhopadhyay SM (2012) Novel platform development using an assembly of carbon nanotube, nanogold and immobilized RNA capture element towards rapid, selective sensing of bacteria. Sensors 12:8135–8144

Perez JM, Simeone FJ, Saek Y, Josephson L, Weissleder R (2003) Viral induced self-assembly of magnetic nanoparticles allows the detection of viral particles in biological media. J Am Chem Soc 125:10192–10193
Rai V, Sefali A, Nrisingha D (2012) Implications of nanobiosensors in agriculture. J Biomater Nanobiotechnol 3:315–324
Nabiabad HS, Piri K, Amini M (2015) Nanobiosensors-their applications in the medicinal plants industry. J Med Plants By-products 2:131–140
Secil C (2008) Development of biosensors for determination of the total antioxidant capacity. Dissertation, Izmir Institute of Technology
Simona CL, Sandra AV, Mirela D, Andreia T, Gabriel Lucian R (2011) Biosensors applications on assessment of reactive oxygen species and antioxidants. J Environ Biosens 9:12–21
Srilatha B (2011) Nanotechnology in agriculture. J Nanomed Nanotechnol 2:7–11
Sumner JP, Aylott JW, Monson E, Kopelman R (2002) A fluorescent PEBBLE nanosensor for intracellular free zinc. Analyst 127:11–16
Turner AP (2000) Biosensors sense and sensitivity. Science 290:1315–1317
Updike SJ, Hicks GP (1967) The enzyme electrode. Nature 214:986–988
Velasco MN (2009) Optical biosensors for probing at the cellular level: a review of recent progress and future prospects. Semin Cell Dev Biol 20:27–33
You C, Bhagawati M, Brecht A, Piehler J (2009) Affinity capturing for targeting proteins into micro and nanostructures. Anal Bioanal Chem 393:1563–1570
Zhang W, Tang H, Geng P, Wang Q, Jin L, Wu Z (2007) Amperometric method for rapid detection of *Escherichia coli* by flow injection analysis using a bismuth nano film modified glassy carbon electrode. Electrochem Commun 9:833–838
Ziegler C (2004) Cantilever based biosensors. Anal Bioanal Chem 379:946–959

Chapter 11
Nanomaterials and Agrowaste

Sumera Javad, Iqra Akhtar, and Shagufta Naz

11.1 Nanotechnology in Agriculture

Nanotechnology is going to be more and more influential in the agriculture with increasing and extensive research in both fields. Modern technologies demand us to reduce the cost with complete risk assessments of the procedures with maximum productions. There are number of recent advances in nanomaterials like fullerenes, nanofibers, quantum dots and CNT (Carbon nanotubes). These materials have specific physical, optical and mechanical properties which are unique and make their applications possible in different areas of agriculture. These can be used in nanosensors with significant applications in soil analysis, pesticides release, water supply and management of other biochemicals (Prameela 2017). These can also be used as nanofertilizers, nanoemulsions for fertilizer coatings etc. Another use of nanotechnology in agriculture may be suggested for the agrowaste management.

11.2 Agrowaste

Agrowaste is a term used to describe the waste material produced during agricultural practices which may be any chemical, pesticide or fertilizer. These materials are usually hazardous in nature and their use must be reduced but they are required in larger amounts to get optimum products from crops. Agrowaste also includes plant parts which are not used as food.

S. Javad (✉) · I. Akhtar
Department of Botany, Lahore College for Women University, Lahore, Pakistan

S. Naz
Department of Biotechnology, Lahore College for Women University, Lahore, Pakistan

S. Javad (ed.), *Nanoagronomy*, https://doi.org/10.1007/978-3-030-41275-3_11

The significance of sustainability, especially in view of the rise in the worldwide population, has been increasingly recognized. This problem is closely related to a circular economy that is based on resource regeneration. Waste reduction is one of the key elements of circular economy. Agricultural waste is defined as "waste generated as the consequence of a variety of agricultural activities, including manure and other waste, from farms, harvesting waste, landfill fertilizers, pesticides entering water, air or soil, and field-drained salts and silt" (Bruce et al. 2005). In the form of residues and waste, a substantial portion of agro food production is lost. Therefore the exploration of innovative techniques that provide new possibilities for complete sustainability is extremely important. Nanotechnology is assumed to contribute considerably in this direction too (Chen et al. 2013).

Each year, a lot of food and agricultural products are being lost world wide as agrowaste. 1/3 of food produced every year is wasted according to one estimate on the name of human consumption (almost 1.6 G tons according to one estimate). Whereas another estimate claims that there is an overall 6 G tons loss of food and non-food agricultural products every year. Firstly, if we reduce this much loss of agricultural products, we can reduce the pressure on our resources. It will also lower the increasing demand of more food availability for ever increasing population, thus automatically reducing the use of chemical fertilizers and pesticides.

When crop is harvested, there are other related issues like crop waste which is almost 80% of the biomass of agriculture. Mostly farmers burn this waste which causes the emission of gasses into the environment in bulk quantities. Now a day, it's the main reason of smog (fog + smoke) and posing a serious threat to health of common man. Therefore it is the need of hour to manage the waste in a strategic way to recover, recycle and reuse the agrowaste (Krishnaswamy et al. 2014).

This Agrowaste, if used properly, can be a part of economy and progress. There are a lot of uses suggested like in production of biotechnological products, synthesis of nanoparticles, manure and biofuel production etc. (Fig. 11.1).

Fortunately, instead of remaining in the environment, different types of reuse methods are proposed for agricultural waste (Tepe and Dursun 2014). The residues which are acquired from agriculture or agricultural products like food, feed and

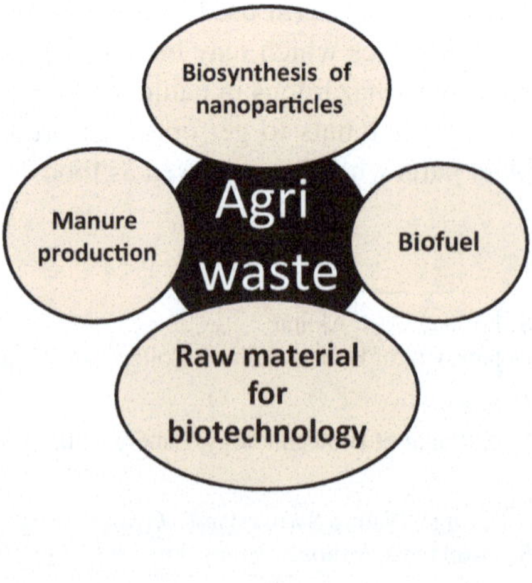

Fig. 11.1 Probable management of agrowaste

fruits etc. contain lignocellulose (Chandra et al. 2012). Therefore these waste materials can be used as a raw material for number of biotechnology products like biofuels, enriched protein, microbial pigments, antibiotics, aromatic compounds, secondary bioactive metabolic products, organic acids, enzymes and nanofibers (Nigam and Pandey 2009).

11.3 Types of Agrowaste

Agrowaste is of different types. Mainly it can be classified into two main categories i.e., Agricultural residues and agricultural industry residues. Agricultural residues are of further two types namely field residues (stems, seeds, stalks etc.) and process residues (Husk, Bagasse, Molasses). Agricultural industry waste may contain any agricultural waste produced at commercial level like orange peel, potato peel, ground nut oil cake and soybean oil cake etc. (Sadh et al. 2018). Waste is largely obtained from coffee, cereal, cocoa, fruit, cotton, tea, vegetables etc. These waste materials can produce organic solvents such as ethanol, acetone and butanol; can also be used to produce biofuels (Carchi 2014). Leaf fiber of pineapple is also being investigated to substitute fiberglass. It is proposed to be used as a construction material particularly (Souza et al. 2010). Wine-producing waste is another example which is a huge representative of lignocellulosic waste. It is produced in such a huge quantity, like in just Spain, every year over 1 million tons residues of wine waste are produced. These residues are being used as cultivation media, land modifications, production of lactic acid, phenolic production, antioxidant compounds and biosurfactants, and low cost adsorbent of heavy metals.

Banana waste produces huge quantities of fake stem, foliar and shells waste that are problematic to degrade and cause environmental pollution (Mejía et al. 2010). Cellulose, present abundantly in nature and from almost every plant organism, is one of the enticing compounds produced from agricultural waste. This compound can be extracted through the large quantities of agricultural waste of banana, and a possible removal of nanocellulose can specifically be attempted. Some plant areas, like the pseudo stem, have been found to have a cellulose content of 38%, with the floral stem representing 27%. The importance of its extraction exists in the probability of substituting certain synthesized fibers which may pollute the environment, taking advantage of excess properties and nanocytes offering the production of certain products of commercial origin such as hemicellulose and lignin with cellulose and other components of plants (Carchi 2014). It has a good property such as a low density as 1.6 g/cm^3 and a responsive surface, which has OH groups. Presence of these OH groups on the surface enable certain chemicals to be attached, which allowing them to acquire new surfaces. This ability is used to modify the configuration and regulation of the particle-particle bond and the particle-matrix. Furthermore, the tensile strength is good, and the thermal expansion coefficient is low. These products can be used as polymer strengthening products, biomedical products, textiles, super capacitors, pharmaceuticals, and antimicrobial films (Nairn et al. 2011).

Cellulose is an Earth's most abundant, fibrous semi-crystalline biopolymer. It is the primary component of plant matrixes. The cell walls of plants consist of rigid microfibrils of cellulose embedded into a matrix of lignin and soft hemicelluloses. Cellulose chains are arranged to produce compact, intermolecular as well as intramolecular hydrogen bonds that are stabilized to form compaction. These microfibrils consist of primary fibers, 8–50 nm in diameter and some microns in length. Nanofibers give strength to the plant stem due to their crystal structure. The biggest challenge to adopt this new technology is to increase the production size and applications of nanofibers in industry. One way to do this is to build synergies among sectors which use natural fibers, such as packing, cars, textiles and medicines, and possible manufacturers of natural nano-fiber (Prameela 2017).

One of the world's major used juices by health-conscious people is orange juice. After extraction, around half the weight of the fruit is transformed into citrus peel waste of orange (CPWO), a low-cost material that is best used for the manufacturing and production of bioethanol of the first generation (Awan et al. 2013). The researchers also proved that, by using a new laboratory-sized direct steam injection device, oranges peel waste can be fermented by the strain *Saccharomeces cerevisiae* which produce large amount of glucose and ethanol levels (up to 50% m^{-1}). Dry orange peel waste is also documented as an effective material for bioethanol production with a total process yield (mass equilibrium) at 140 L bench reactor scale (Santi et al. 2012). CPWO is also a valuable bioresource for obtaining D-limonene (1.5% g^{-1} of CPWO), nanocellulose (3% g^{-1} of dry CPWO), and bioethanol (20% g^{-1} of dry CPWO), which has been determined on a dry-material basis in addition to bioethanol production (Tsukamoto et al. 2013).

The use of pineapple biomass has become interesting in tropical countries due to its ecological properties and renewable nature. Pineapple leaf is currently a waste of the cultivation of pineapple. These fibers can be acquired without the need for extra costs from industrial sectors. The composition comprises of a vascular bundle system that is made of bundles of fibrous cells, obtained after all epidermis tissues have been mechanically removed. The fibers are, on the other side, highly hygroscopic and comparatively cheap. The top mechanical properties of pineapple leaf fiber have been determined to be multicellular and lignocellulosic because of the high cellulos, the low microfrilliary angle of 14°. The distinctive characteristics shown by an apple leaf may be applied in composite matrices as possible enhancement and will be considered to substitute glass fiber in low-cost products in particular construction materials (Bakshi et al. 2014).

11.4 Nanomaterial and Agrowaste

Agrowaste can be treated using nanoscale devices in various ways. There are multiple examples like nanoscale iron oxide, other metal nano-oxides and nanotubes etc. which are being used in the farmland waste water treatment. Nano-electron enzymes are another source which are being used for water remediation. Clay nano-mineral hydrotalcite is the technology which can be used for water treatment along with porous candles (Gillman 2006). There are lots of examples where industrial waste has provided a base for replacement of fossil fuels and also acted as a source for the manufacturing of nanomaterials. In the same way, biological waste and particularly, Agrowaste can play its own role if used in a proper and environment friendly way. It can be used to prepare biocomposites which are more sustainable in production (Prasad et al. 2017).

Another way may be the removal of these wastes from environment by using nanoparticles or nanoformulations. For example, pesticides like Malathion and Chlorpyrifos can be removed from air by using nanoparticles. There are different types of nanofilters available in various countries for spot maintenance (Karn et al. 2009). Titanium oxide nanoparticles are used with filters to remove photocatalytic contaminants from water (McMurray et al. 2006). Wheat straws and soy fibers are also being used as source for manufacture of bio-nanocomposites.

Some of the significant uses of Agrowaste can be recommended as follows.

11.4.1 Biosynthesis of Nanoparticles by Post-harvest Agrowaste

The concept of green nanotechnology is currently being developed by scientists. This integrates green chemistry and green technology principles to generate green, secure and non-poisonous nanoparticles in the synthesis protocol (Llantén et al. 2013). Based on their unique characteristics it is therefore of great interest that nanoparticles of noble metals, like gold nanoparticles (AuNP), be produced. Semicontractor, spectroscopy, biomedical applications like medicinal products, imaging of the tissues and tumors, and cancer treatment are all applications (Seeram et al. 2005). Nanoparticles can be synthesized by using plants extracts in a cost effective way which has been described earlier. But, nanoparticles can also be synthesized by using post-harvest agrowaste as a raw material instead of edible part of plants. Various agrowaste employed in green synthesis of nanoparticles reported in literature is presented in Table 11.1.

All the factors involved in controlling the shape and size of nanoparticles are the same as are those controlled for the synthesis of nanoparticles from plant extract i.e., light, temperature, pH and time of reaction etc. These factors can be optimized for each type of Agrowaste and type of nanoparticles by using suitable experimental designs (Edmundson et al. 2014). The formation of metal nanoparticles of silver, copper, gold, titanium oxide, zinc oxide can be characterized by using the techniques like Transmission electron microscopy (TEM), X-ray diffraction (XRD) analysis, Scanning electron microscopy (SEM), UV-visible spectroscopy etc.

New reports have demonstrated that water is increasingly used as solvent in chemical reactions rather than organic solvents. The water-dispersible nature of

Table 11.1 Synthesis of nanoparticle using agro-waste

Sr. No	Agrowaste	Nanoparticles	References
1	Peels of Pomegranate	Silver nanoparticles	Ahmad and Sharma (2012)
2	Coconut shells	Silver nanoparticles	Sinsinwar et al. (2018)
3	*Psidium guajava* waste	Silver nanoparticles	Bose and Chatterjee (2016)
4	Sugarcane waste	Silver and Gold nanoparticles	Mishra and Sardar (2016)
5	Macadamia nut shells	Gold nanoparticles	Dang et al. (2019)
6	Peels of Banana	Gold nanoparticles	Vijayakumar (2017)
7	Wastes of grapes	Gold nanoparticles	Krishnaswamy et al. (2014)

grape-by-product gold nanoparticles together with their health benefits can revolutionize the effect of nanotechnology on the supply of medicines.

These formed nanoparticles form the basis for production of a number of nanoparticles structures like nanodevices, nanosensors, nanocatalysts and nanomedicines (Kulkarni 2015).

The agricultural waste may be used in nanomaterial synthesis, but a lot of work is required to adopt those protocols and techniques of synthesis for their standardization. The use of agricultural waste for the manufacturing of nanoparticles would contribute to sustainable development for the environment and also for the development of the economy.

11.4.2 Biosynthesis of NPs by Using Weeds

Weeds in a field are the undesirable plants, herbs, or shrubs. Usually, weeds are demolished and burnt. Although many weeds have demonstrated significant pharmaceutical properties and are further investigated. The capability of weeds for nanomaterial synthesis and as bio-reactors cannot be rejected, and nanoparticles should also be more and more explored because weeds transmitted in a green synthesis do not require the disposal of large green trees and would be more convenient and suitable for the formation of nanoparticles (Zamar et al. 2016). The Table 11.2 below summarizes the recent studies on green synthesis of nanoparticle mediated through weeds.

11.4.3 Bio-nanocomposites from Agricultural Residues

Since the beginning of the twentieth century, natural vegetation and wood fibers have been utilized as refinement materials for bio-composite production. These biomedical composites are of interest as a substitute for reinforced composites made from synthetic fibers in increasing numbers of industries including the automotive, packaging, construction and consumer products industry, with their relative strength,

Table 11.2 Weeds used for synthesis of nanoparticle

Sr. No	Weeds used	Nanoparticles	References
1	*Cyperus rotundus*	Silver Nanoprticles	Syafiuddin et al. (2017)
2	*Clidemia hirta*	Silver Nanoparticles	Trujillo et al. (1986)
3	*Medicago sativa*	Silver Nanoparticles	Roy and Barik (2010)
4	*Gloriosa superba*	Copper Nanoparticles	Pawar et al. (2016)
5	*Tinospora cordifolia*	Gold Nanoparticles	Abbasi et al. (2014)

high rigidity, low density and being renewable. In applications like feedstock and energy production, only a small percentage is used.

11.4.4 Cellulose Nanofibers

The most important components of the plant tissues are cellulose, which is the most common biopolymer on Earth. The most important industrial source of cellulose is the lignocellulosic material that is present in wood. Other materials containing cellulose include farm waste, water plants, grasses and other plant materials. The worldwide amount of cellulose produced by photosynthesis is estimated between 1011 and 1012 tons of cellulose per year (Wang et al. 2015). The development units of the hierarchical system of cellulose fibers represent micro and macro fiber in plant tissues. In its turn, microfibrils comprise of elementary (nanofiber) fibrils which (depending on cellulose sources) have a diameter of about 3–35 nm (Athinarayanan et al. 2015). In recent years, nanocellulose attracted considerable attention as a new bio-based nanomaterial with excellent optical properties, great strength and a particular surface area (Wanyika et al. 2012). For a broad range of nanocomposite applications nanocellulose can be extracted from wood waste and can be chemically modified (Shen 2017). Different farm plants, including residues like soy and wheat straw, sugar beet pulp, potato pulp and rutabaga, have already been regarded raw materials for new cost-effective manufacturing of nanocellulose (Shankar et al. 2016).

11.4.5 Rice Husk-Derived Si Nanomaterials

In 2017, 503.8 million tons for the worldwide paddy production was estimated by FAO (milled basis) (Bhuyan et al. 2016). Almost 25% of this manufacturing is made from rice husk (RH), discarded as a rice milling by-product. The rice husk is the layer on a rice grain that protects the seed during the cultivation process. Rice husks consist mainly of lignocellulose (approximately 72–85 wt%) and silica (approx. 15–28 wt%) (Muramatsu et al. 2014). The second most important component of the Earth's crust is silicon. Grasses absorb excess quantities of Si and place them on the plant as amorphous, hydrated silica throughout their life cycle ($SO_2 \cdot nH_2O$). The Si content in grass ash can be between 50% and 70% (Somanathan et al. 2015).

11.4.6 Orange Peel Cellulose and Nanocellulose in Textile and Food Industry

Orange juices are manufactured at an industrial level of large scale, which is still considered to be sufficient agricultural waste in a considerable quantity of solid and liquid residues (around 8–20 million tons per year). Orange residues normally have no commercial value but the residue includes rich, soluble, insoluble carbohydrates, pectin, essential oils, cellulose and hemicellulose which are the basis of various industries (Awan et al. 2013; Tsukamoto et al. 2013). The orange nanofibril production from these residues therefore not only reduces costs but also plays a major role in the development of qualitative and sustainable textiles. Citrus by-products are diverse for the recovery of environmentally sound nanofibers with greater capacities for water and oil retention, great fermentation capacity and lower calorie content used for functional ingredients for human health (Lario et al. 2004; Fernando et al. 2005).

11.4.7 Graphene Oxide

The possible use of graphene oxide (GO) has been observed in various fields in science. It is single layer of graphite and may be insulator or conductor. GO synthesis from sugar cane, bagasse and rice husk ash has proved its importance and cost effectiveness. The available GO synthesis protocols from agro-waste are however already in its early stages and further development is going on in modern methods to synthesize eco-friendly, economical, agro-waste GO compounds (Somanathan et al. 2015).

11.5 Future Prospective

Research into biodegradable and biocompatible materials with the concept "green" materials is increasingly pressing for environmental sustainability. In this context, nano-biomaterial from agriculture waste may be considered an attractive product having biodegradable, removable or biocompatible propensities. These products may have an excellent thermal stability in addition to being mechanically strong, stiff and very crystalline. Nanocellulose is a major pollution management player. It is said so due to the potential role of nanocellulose for biomass degradable and its broad applications in particular to the environmental concerns. The extraction of nanocellulose from agriculture waste, such as citrus and orange, seems as promising waste treatment replacements. The improved crop yields by reducing the use of fertilizer and pesticides (SiO_2 NPs) were one of the major compounds for improving farm yields. As promising instruments for the production of carbon nanoparticles, agriculture waste could be utilized to obtain the best possible return from agricultural waste. Despite these potential points of interest, the synthesis of bio-nanomaterial from agricultural waste its own importance. Research is going on to accomplish its limits to the market sector to a certain degree. In the future, therefore, there should be positive aspects for the marketing of these extremely wealthy large-scale bio-nanomaterials. New research is also intended to improve the efficiency of use of water for plants, pesticides and fertilizers, pollution reduction and environmentally friendly agriculture.

References

Abbasi T, Anuradha J, Abbasi SA (2014) Utilization of the terrestrial weed guduchi (*Tinospora cordifolia*) in clean-green synthesis of gold nanoparticles. Nanosci Tech 1(3):1–7

Ahmad N, Sharma S (2012) Green synthesis of silver nanoparticles using extracts of *Ananas comosus*. Green Sustain Chem 2:141–147

Athinarayanan J, Periasamy VS, Alhazmi M, Alatiah KA, Alshatwi AA (2015) Synthesis of biogenic silica nanoparticles from rice husks for biomedical applications. Ceram Int 41(1):275–281

Awan AR, Manfredo A, Pleiss JA (2013) Lariat sequencing in a unicellular yeast identifies regulated alternative splicing of exons that are evolutionarily conserved with humans. Proc Natl Acad Sci 110:12762–12767

Bakshi S, Choi H, Weisshaar JC (2014) The spatial biology of transcription and translation in rapidly growing *Escherichia coli*. Front Microbiol 6:636. https://doi.org/10.3389/fmicb.2015.00636

Bhuyan DJ, Vuong QV, Chalmers AC, Altena BA, Bowyer MC, Scarlet CJ (2016) Investigation of phytochemicals and antioxidant capacity of selected *Eucalyptus* species using conventional extraction. Chem Pap 70(5):567–575

Bose D, Chatterjee S (2016) Biogenic synthesis of silver nanoparticles using guava (*Psidium guajava*) leaf extract and its antibacterial activity against *Pseudomonas aeruginosa*. Appl Nanosci 6(6):895–901

Bruce PG, Armstrong G, Armstrong AR, Canalesa J (2005) Nanotubes with the TiO_2-B structure. Chem Commun 19:2454–2456

Carchi M (2014) Utilization of agricultural residues derived from the cultivation of bananas for nanocellulose. Carbohydr Polym 81(3):720–725

Chandra R, Takeuchi H, Hasegawa T (2012) Methane production from lignocellulosic agricultural crop wastes: a review in context to second generation of biofuel production. Renew Sustain Energy Rev 16(3):1462–1476

Chen H, Wang B, Gao D, Guan M, Zheng L, Ouyang H, Chai Z, Zhao Y, Feng W (2013) Broad-spectrum antibacterial activity of carbon nanotubes to human gut bacteria. Small 9(16):2735–2746

Dang H, Fawcett D, Poinern GEJ (2019) Green synthesis of gold nanoparticles from waste macadamia nut shells and their antimicrobial activity against *Escherichia coli* and *Staphylococcus epidermis*. Int J Res Med Sci 7(4):1171–1177

Edmundson MC, Capeness M, Horsfall L (2014) Exploring the potential of metallic nanoparticles within synthetic biology. N Biotechnol 25(31):572–578

Fernando F, Maria LH, Ana ME, Italo C, Fernando A (2005) Fiber concentrates from apple pomace and citrus peel as potential fiber sources for food enrichment. Food Chem 91:395–401
Gillman GP (2006) A simple technology for arsenic removal from drinking water using hydrotalcite. Sci Total Environ 366(2–3):926–931
Karn B, Kuiken T, Otto M (2009) Nanotechnology and *in situ* remediation: a review of the benefits and potential risks. Environ Health Perspect 117(12):1813–1831
Krishnaswamy K, Valib H, Orsata V (2014) Value-adding to grape waste: green synthesis of gold nanoparticles. J Food Eng 142:210–220
Kulkarni SK (2015) Nanotechnology: principles and practices, 3rd edn. Springer, New Delhi
Lario Y, Sendra E, Garci J, Perez A, Fuentes C, Barbera ES, Lopez JF, Alvarez JAP (2004) Preparation of high dietary fiber powder from lemon juice by-products. Innov Food Sci Emerg Technol 5(1):113–117
Llantén DNC, Munoz SA, Castro ME, Muñoz PA, Blamey JM (2013) Gold nanoparticles synthesized by Geobacillus sp. strain ID17 a thermophilic bacterium isolated from Deception Island, Antarctica. Microb Cell Fact 12:75. https://doi.org/10.1186/1475-2859-12-75
McMurray TA, Dunlop PSM, Byrne JA (2006) The photocatalytic degradation of atrazine on nanoparticulate TiO_2 films. J Photochem Photobio A Chem 182:43–51
Mejía HE, Quintana GC, Ogunsile BO (2010) Development of binderless fiber boards from steam-exploded and oxidized oil palm wastes. Biol Res 9(2):2922–2936
Mishra A, Sardar M (2016) Rapid biosynthesis of silver nanoparticles using sugarcane bagasse-an industrial waste. J Nanoeng Nanomanufact 3:1–3
Muramatsu H, Kim YA, Yang KS, Cruz-Silva R, Toda I, Yamada T, Terrones M, Endo M, Hayashi T, Saitoh H (2014) Rice husk-derived graphene with nano-sized domains and clean edges. Small 10:2766–2770
Nairn J, Moon RJ, Martini A, Simonsenf J, Youngblood J (2011) Cellulose nanomaterials review: structure, properties and nanocomposites. Chem Soc Rev 40(7):3941–3994
Nigam P, Pandey A (2009) Biotechnology for agro-industrial residues utilization: utilization of agro-residues. Springer, New York, NY. https://doi.org/10.1007/978-1-4020-9942-7
Pawar O, Deshpande N, Dagade S, Waghmode S, Joshi NP (2016) Green synthesis of silver nanoparticles from purple acid phosphatase apoenzyme isolated from a new source *Limonia acidissima*. J Exp Nanosci 11:28–37
Prameela LK (2017) Nanomaterial's applications in agriculture. J Chem Pharm Sci 10(1):593–596
Prasad R, Kumar M, Kumar V (2017) Nanotecnology: an agriculture paradigm, vol 10. Springer Nature, Singapore, pp 978–981
Roy N, Barik A (2010) Green synthesis of silver nanoparticles from the unexploited weed resources. Int J Nanotech Appl 4(2):95–101
Sadh PK, Duhan S, Duhan JS (2018) Agro-industrial wastes and their utilization using solid state fermentation: a review. Bioresour Bioprocess 5:1. https://doi.org/10.1186/s40643-017-0187-z
Santi G, Crognale S, Moresi M, Petruccioli M, Annibale A (2012) Improved orange peel waste pretreatments for bioethanol production. Environ Eng Manag J 11(3):55
Seeram NP, Adams LS, Henning SM (2005) *In vitro* antiproliferative, apoptotic and antioxidant activities of punicalagin, ellagic acid and a total pomegranate tannin extract are enhanced in combination with other polyphenols as found in pomegranate juice. J Nutr Biochem 16(6):360–367
Shankar S, Wang LF, Rhim JW (2016) Preparations and characterization of alginate/silver composite films: effect of types of silver particles. Carbohydr Polym 146:208–216
Shen Y (2017) Polymer nanocomposite dielectrics for electrical energy storage. Natl Sci Rev 4(1):23–25
Sinsinwar S, Sarkar MK, Suriya KR, Nithyanand P, Vadivel V (2018) Use of agricultural waste (coconut shell) for the synthesis of silver nanoparticles and evaluation of their antibacterial activity against selected human pathogens. Microb Pathog 124:30–37
Somanathan T, Prasad K, Ostrikov K, Saravanan A, Krishna VM (2015) Graphene oxide synthesis from agro waste. Nanomaterials 5(2):826–834

Souza CPF, Souza EH, Ledo CAS, Souza FVD (2010) Evaluation of the micropropagation potential of curauá pineapple hybrids for fiber production. Acta Amazon 48(4):290. https://doi.org/10.1590/1809-4392201800382

Syafiuddin A, Salmiati S, Salim MR, Kueh ABH, Tony H, Hadi N (2017) A review of silver nanoparticles: research trends, global consumption, synthesis, properties and future challenges. J Chin Chem Soc 64:732. https://doi.org/10.1002/jccs.201700067

Tepe Ö, Dursun AY (2014) Exo-pectinase production by *Bacillus pumilus* using different agricultural wastes and optimizing of medium components using response surface methodology. Environ Sci Pollut Res 21(16):9911–9920

Trujillo EE, Latterell F, Rossi A (1986) *Colletotrichum gloeosporioides*, a possible biological control agent for *Clidemia hirta* in Hawaiian forests. Plant Dis 70:974–976

Tsukamoto J, Duránb N, Tasic L (2013) Nanocellulose and bioethanol production from orange waste using isolated microorganisms. J Braz Chem Soc 24(9):1537–1543

Vijayakumar S, Vaseeharan B, Malaikozhundan B, et al (2017) Therapeutic effects of gold nanoparticles synthesized using Musa paradisiaca peel extract against multiple antibiotic resistant Enterococcus facealis biofilms and human lung cancer cells (A549). Microb Path. 102:173–183.

Wang W, Mozuch MD, Sabo RC, Kersten P, Zhu JY, Jin Y (2015) Production of cellulose nanofibrils from bleached eucalyptus fibers by hyperthermostable endoglucanase treatment and subsequent microfluidization. Cellul 22:351–361

Wanyika H, Gatebe E, Kioni P, Tang Z, Gao Y (2012) Mesoporous silica nanoparticles carrier for urea: potential applications in agrochemical delivery systems. J Nanosci Nanotechnol 12(3):2221–2228

Zamar D, Vutukuru SS, Babu R (2016) Biosynthesis of nanoparticles from agro-waste: a sustainable approach. Int J Eng App Sci Tech 1(12):85–92

Chapter 12
Prospects and Constraints

Sumera Javad and Aneeqa Sabah Nazir

12.1 Introduction

Nanotechnology has gained importance in several fields of industry and agriculture. Nanoparticles, due to their unique chemical and physical characters have proven to be useful in electronics, material sciences, energy sector, and pharmaceutical industry and in agriculture as well (Parisi et al. 2015). One of the important characters of nanoparticles is their sustainable competitiveness.

12.2 Agronomy

Agronomy is the basic activity of human race to survive on the face of earth (Reddy 2015). Agricultural systems are mainly of two types,

1. Conventional and
2. Conservation or sustainable agriculture.

First type includes conventional methods of farming with primitive type of preventive measures. These measures may include crop rotation and surface crop residue retention. These methods are able to get many goals so far but they are disturbing the nature at a faster pace (Farooq and Siddique 2015). Second type of agriculture is more planned version of farming with well-planned use of resources along with the reduced use of sprays or chemicals.

S. Javad (✉)
Department of Botany, Lahore College for Women University, Lahore, Pakistan

A. S. Nazir
Department of Physics, Lahore College for Women University, Lahore, Pakistan

S. Javad (ed.), *Nanoagronomy*, https://doi.org/10.1007/978-3-030-41275-3_12

Fig. 12.1 Inter-relationship of driving forces of nature and society

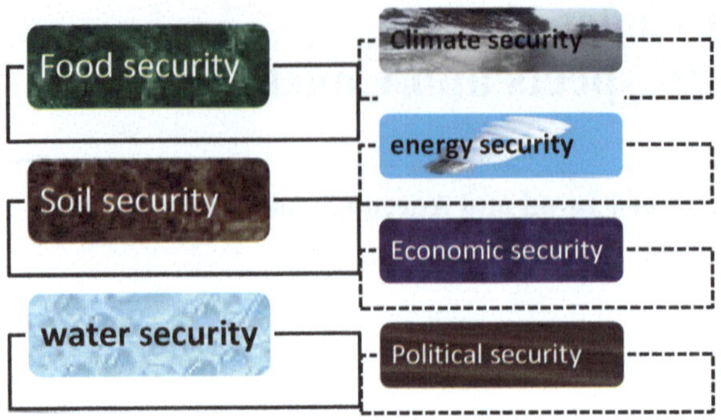

Agriculture is the area of research where innovations are continuously needed to face the scarcity of food, water and agricultural land with every passing day. Nanotechnology has very direct applications in processing and packaging of foods but its future directly related to agronomy is still uncertain (Parisi et al. 2015).

Scientists have mentioned about the inter-relationship of different driving forces of world which are mentioned in Fig. 12.1.

For example food security includes its availability, nutritional level and quality etc., which is related to water and soil quality of an area. In the same way economic stability depends upon food security and makes the shape of political security of a country and nation (Lal 2015). Therefore these natural resources are said to be in close connection with the anthropogenic factors, giving shape to the future of a society. It is the need of the hour to optimize the conditions to develop positive links and connections of these forces. Therefore much progress in this regard is required in plant nutrition and idea of nano-nutrition for sustainable agriculture production (Ditta et al. 2015; Shalaby et al. 2016).

12.3 Associated Issues

Nanotechnology is known for its advantages for its applications in the field of agriculture, but they are yet considered as marginal by a number of scientists and policy makers. Already techniques and tools are available for agricultural purposes (Fig. 12.2). Nano-techniques have not intruded so much in the market of agriculture as the traditional industries are. Research discoveries are going on in the field of nano-agronomy but they are still at the level of academics. Industries are not taking chances.

Another associated problem is that the stakeholders of big traditional industries have big patents to show. It is so common in traditional agro-chemical companies and their number of patents is growing on. This should be the main concern because these big industries are free to play solo games with freedom in future to operate the commercial developments according to their benefits.

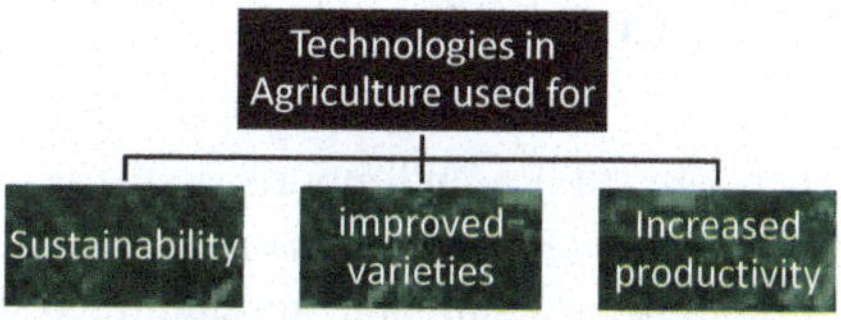

Fig. 12.2 Earlier use of technology in agriculture

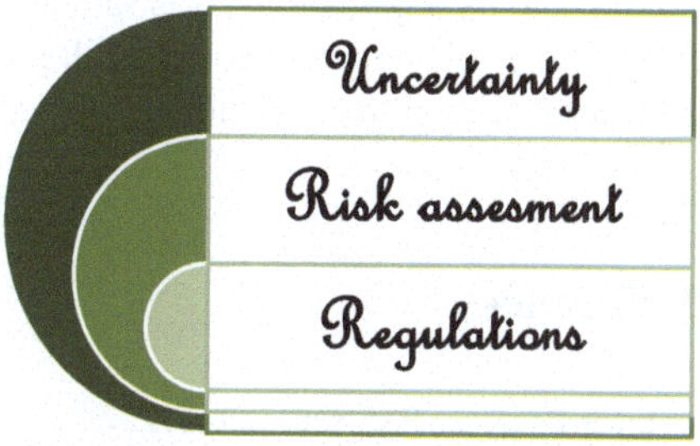

Fig. 12.3 Problems associated with nanotechnology in agronomy

Different regulatory bodies are also working on nanotechnology research and products for agronomy. They have to confirm the quality assurance and risk assessment. They are pressing hard to mention nano—on the label which may affect the public opinion. A number of products already available in market are consisting of nanosized materials involuntarily. They include clay, silica etc. A common man usually hesitates to accept new technologies and according to one school of thought it is the main cause of rejection of nano-products in the market. Whereas another school of thought also explains that over all opinion of public is not against the nanotechnology (Caliman et al. 2011). Uncertainty and availability are the two main factors which are governing the situation. Commercialization is the main step which can help people to perceive the idea of benefits and advantages of nanotechnology in agronomy. Consequently, it could be established that, nanomaterials or nanoparticles can be used in sustaining the agricultural sectors together with water, energy, food and soils within the limits of water security, soil security, food security and energy security. There is a need to broaden our views to consider the potential of nanoparticles and nanotechnology in the field of agronomy for the sake of broader benefits (Shalaby et al. 2016). There is also a great need to resolve the issue of engineered nanomaterials.

There is much literature available about the potential use of nanotechnology to revolutionize the agronomy worldwide. Nanotechnology offers novel tools and technologies to enhance the agricultural production by providing nanofertilizers, nanozeolites, nanoherbicides, nanopesticides and their efficient and targeted applications. Slow and targeted release of pesticides and water to the crops, enhanced germination, in-time disease detection and management are the advantages related to nanotechnology and can attract non-scientific consumer to make things economical as well. When these technologies are going to solve the limitations and challenges of intensive traditional farming system (Fig. 12.3) then people will be definitely accepting it (Ditta et al. 2015; Tarafdar et al. 2013; Scrinis and Lyons 2007).

12.4 Conclusion

There are studies and reports on the positive impacts of nanoparticles on germination and growth of various crops. It is believed that nanoparticles have a great role to establish a sustainable agriculture system. Reports related to effect of nanoparticles on crop germination are very strong. Future of agronomy lies in the nanotechnology and its applications. It is said and well said "While it may be hard to predict what future role nanotechnology will play in the development of fertilizers, there is a clear indication that the industry is heading in this direction" (DeRosa 2009).

References

Caliman FA, Robu BM, Smaranda C, Pavel VL, Gavrilescu M (2011) Soil and groundwater cleanup: benefits and limits of emerging technologies. Clean Technol Environ Policy 13(2):241–268

DeRosa M (2009) Prospects and potential impacts of nanotechnology on fertilizer inputs: a foresight review on the use of nanotechnology in agriculture. Can Food Inspect Agency 31:7

Ditta A, Arshad M, Ibrahim M (2015) Nanoparticles in sustainable agricultural crop production: applications and perspectives. In: Siddiqui MH, Al-Whaibi MH, Mohammad F (eds) Nanotechnology and plant sciences: nanoparticles and their impact on plants. Springer International Publishing, Cham, pp 55–75. https://doi.org/10.1007/978-3-319-14502-4

Farooq M, Siddique KHM (2015) Conservation agriculture. Springer International Publishing, Cham. https://doi.org/10.1007/978-3-319-11620-4

Lal R (2015) The nexus approach to managing water, soil and waste under changing climate and growing demands on natural resources. In: Kurian M, Ardakanian R (eds) Governing the nexus: water, soil and waste resources considering global change. Springer, Cham, pp 39–61. https://doi.org/10.1007/978-3-319-05747-7_3

Parisi C, Mauro V, Emilio R (2015) Agricultural nanotechnologies: what are the current possibilities? Nano Today 10:124–127

Reddy PP (2015) Climate resilient agriculture for ensuring food security. Springer, New Delhi. https://doi.org/10.1007/978-81-322-2199-9

Scrinis G, Lyons K (2007) The emerging nano-corporate paradigm: nanotechnology and the transformation of nature, food and agri-food systems. J Soc Food Agric 15(2):22–44. ISSN: 0798–1759

Shalaby TA, Yousry B, Abdalla N, Taha H, Tarek A, Said S, Megahed A, Eva D, Hassan ER (2016) Chapter 10. Nanoparticles, soil, plants and sustainable agriculture. In: Shivendu R, Nandita D, Eric L (eds) Nanoscience in food and agriculture, vol 1. Springer, Cham, pp 283–312. https://doi.org/10.1007/978-3-319-39303-2_10

Tarafdar JC, Sharma S, Raliya R (2013) Nanotechnology: interdisciplinary science of applications. Afr J Biotechnol 12(3):219–226

Index

S. Javad (ed.), *Nanoagronomy*, https://doi.org/10.1007/978-3-030-41275-3

O

P

Q